T0254785

Signals and Communication Technology

More information about this series at http://www.springer.com/series/4748

Orhan Gazi

Information Theory for Electrical Engineers

Orhan Gazi
Department of Electronics
and Communication Engineering
Çankaya University
Ankara
Turkey

ISSN 1860-4862 ISSN 1860-4870 (electronic)
Signals and Communication Technology
ISBN 978-981-13-4149-6 ISBN 978-981-10-8432-4 (eBook)
https://doi.org/10.1007/978-981-10-8432-4

Softcover re-print of the Hardcover 1st edition 2018

Printed on acid-free paper

This Springer imprint is published by Springer Nature
The registered company is Springer Nature Singapore Pte Ltd.
The registered company address is: 152 Beach Road, #21-01/04 Gateway East, Singapore 189721, Singapore

Preface

Information is a phenomenon that has meaning in human brain. Almost 70 years ago Shannon published his paper in which he defined fundamental mathematical concepts to identify and measure information. Since then, a huge improvement has occurred in communication technology. It is very important to have knowledge on the fundamental concepts of information theory to understand the modern communication technologies. This book has been written especially for electrical and communication engineers working on communication subject. To comprehend the topics included in the book, it is very essential that the reader has the fundamental knowledge on probability and random variables; otherwise, it will be almost impossible to understand the topics explained in this book.

Although this book has been written for graduate courses, any interested person can also read and benefit from the book. We paid attention to the understandability of the topics explained in this book, and for this reason, we presented the parts in details paying attention to the use of simple and detailed mathematics. We tried to provide detailed solved examples as many as we can. The book consists of four chapters. In Chap. 1, we tried to explain the entropy and mutual information concept for discrete random variables. We advise to the reader to study the concepts very well provided in Chap. 1 before proceeding to the other chapters.

In Chap. 2, entropy and mutual information concept for continuous random variables are explained along with the channel capacity. Chapter 3 is devoted to the typical sequences and data compression topic. In many information theory books, the channel coding theorem is explained as a section of a chapter with a few pages. However, one of the most important discoveries of the Shannon is the channel coding theorem, and it is very critical for the electrical and communication engineers to comprehend the channel coding theorem very well. For this reason, channel coding theorem is explained in a separate chapter, i.e., explained in Chap. 4, in details. We tried to provide original examples that illustrate the concept of rate and capacity achievability in Chap. 4. Since this is the first edition of the book, we just included very fundamental concepts in the book. In our future editions, we are planning to increase the content of the book considering the recent modern communication technologies.

As a last word, I dedicate this book to my lovely daughter "Vera GAZİ" who was four years old when this book was being written. Her love was always a motivating factor for my studies.

Maltepe-Ankara, Turkey
September 2017

Orhan Gazi

Contents

Chapter 1
Concept of Information, Discrete Entropy and Mutual Information

In this chapter, we will try to explain the concept of information, discrete entropy and mutual information in details. To master on the information theory subjects, the reader should have a knowledge of probability and random variables. For this reason, we suggest to the reader to review the probability and random variables topics before studying the information theory subjects. Continuous entropy and continuous mutual information are very closely related to discrete entropy and discrete mutual information. For this reason, the reader should try to understand very well the fundamental concepts explained in this chapter, then proceed with the other chapters of the book.

1.1 The Meaning of Information

Let's first consider the following questions. What is information? How do we measure the information?

Information is a phenomenon that has meaning in human brain. The brain has the capability of interpreting events, concepts and objects and can evaluate these items considering their information content. So, information is a part of the universal architecture, and has a meaning for living creatures.

Now, let's consider the second question. How do we measure the information? To answer this question let's consider some situations and try to think their information contents. Consider that your close friend won a lottery. Since winning a lottery can happen with a very low probability, this event includes significant information for you. And consider that someone came to you and said that today the sun raised from the east. Sun always raises from the east. There is nothing surprising in it. And the probability of sunrise from the east is definitely '1'. Hence, sunrise from the east does not give any information to you.

Let's give several more examples. Assume that you are in the north pole and recording the environment with a camera. Later on when you watch the film you

O. Gazi, *Information Theory for Electrical Engineers*, Signals and Communication Technology, https://doi.org/10.1007/978-981-10-8432-4_1

recorded, you see that there is a large red object on one iceberg. This surprises you and you show interest on the object. Assume that you are a medical doctor and you are inspecting a liver X-ray film. On the film, you see something unusual never seen before. You wonder a lot about the reason for its existence. This unusual thing gives you some information about diagnosis.

From all these examples, we feel that information content of an event is somehow related to the surprise amount of the event that gives to you when it occurs. And you are surprised a lot when something unexpected occurs or something with low occurrence probability takes place. So, we can conclude that the information is inversely related to the probability of occurrence of an event. If the probability of occurrence of an event is very low, when the event occurs, it provides you with too much information. In other words, the occurrence of a less probable event gives you more information.

In human communication, information content of an event is either expressed by speech, or by written papers or by pictures. Speech is nothing but a concatenation of the letters which form words, and words form sentences. Some letters may appear more frequently than others, similarly some words may occur more frequently than others, and this is similar for sentence and paragraphs also. Written materials also are nothing but sequence of letters or words or sentences.

When we deliver a speech, our mouth can be considered as the information source. Since it delivers the speech and the speech has some information content, and the words come out randomly from our mouth. Hence, we can consider an information source as a random variable.

Random variables can be classified as discrete and continuous random variables. A random variable produces values, and considering distinct values we can make a set which is called range set of the random variable. Discrete random variables have countable number of values in their range sets. On the other hand, continuous random variables have uncountable number of values in their range sets.

Assume that $\tilde{X}$ denotes a discrete random variable (RV). The range set of $\tilde{X}$ is denoted by $R_{\tilde{X}}$ and let x_i be symbol such that $x_i \in R_{\tilde{X}}$. For the discrete random variable $\tilde{X}$, the probability mass function $p(x)$ is defined as

$$p(x) = Prob(\tilde{X} = x) \tag{1.1}$$

which gives the probability of the discrete RV producing the symbol x. The probability of symbol x_i in this case is denoted as $p(x_i)$. Since information is inversely proportional to the probability of occurrence, then information content of the symbol x_i can be measured using

$$I(x_i) = \frac{1}{p(x_i)} \tag{1.2}$$

which is a unitless quantity. However, in (1.2) we see that when $p(x_i)$ gets very small values, $I(x_i)$ gets very large values, and this is not a good measurement method. So, it is logical to use a function of $1/p(x_i)$ as in

$$I(x_i) = F\left(\frac{1}{p(x_i)}\right). \tag{1.3}$$

How to choose the $F(\cdot)$ function in (1.3) then? Since we don't want $I(x_i)$ to get very large values for small $p(x_i)$ values, we can choose $F(\cdot)$ as $\log_b(\cdot)$ function. In this case (1.3) turns out to be

$$I(x_i) = \log_b\left(\frac{1}{p(x_i)}\right) \tag{1.4}$$

which is again a unitless quantity, and its base is not determined yet. For the base of the logarithm, any value can be chosen, however, in digital communication since bits are used for transmission, the base of the logarithm in (1.4) can be chosen as '2'. And although (1.4) is a unitless quantity we can assign a unit for it. Since in digital communication information content of an event is transmitted using bit sequences, we can assign 'bits' as unit of measure for (1.4). Hence, the Eq. (1.4) gets its final form as

$$I(x_i) = \log_2\left(\frac{1}{p(x_i)}\right) \text{ bits.} \tag{1.5}$$

Thus, we obtained a mathematical formula which can be used for measuring the information content of a symbol or an event. Shannon almost 70 years ago considered measuring the information content of a symbol and proposed (1.5) in literature. Since then, many advances have been done in communication technology.

1.2 Review of Discrete Random Variables

We said that information source can be considered as a random variable, and random variables are divided into two main categories, which are discrete and continuous random variables. Let's first consider the information source as a discrete random variable. Since it is easier to understand some fundamental concepts working on discrete random variables rather than working on continuous random variables. Let's review some fundamental definitions for discrete random variables. If $\tilde{X}$ is a discrete random variable with probability mass function defined as in (1.1), the probabilistic average (mean) value of $\tilde{X}$ is calculated as

$$E(\tilde{X}) = \sum_x xp(x). \tag{1.6}$$

And for a function of $\tilde{X}$, i.e., $g(\tilde{X})$, the mean value, i.e., probabilistic average value is calculated as

$$E(g(\tilde{X})) = \sum_x g(x)p(x). \tag{1.7}$$

The variance of discrete random variable $\tilde{X}$ is calculated as

$$Var(\tilde{X}) = E(\tilde{X}^2) - [E(\tilde{X})]^2 \tag{1.8}$$

where $E(\tilde{X}^2)$ is evaluated as

$$E(\tilde{X}^2) = \sum_x x^2 p(x). \tag{1.9}$$

Example 1.1 For the discrete R.V. $\tilde{X}$, the range set is $R_{\tilde{X}} = \{x_1, x_2, x_3\}$, and the probability mass function for the elements of the range set is defined as

$$p(x_1) = p_1 \qquad p(x_2) = p_2 \qquad p(x_3) = p_3.$$

Find $E(\tilde{X})$, $E(\tilde{X}^2)$ and $Var(\tilde{X})$.

Solution 1.1 Using (1.6), mean value can be calculated as

$$E(\tilde{X}) = \sum_x xp(x) \rightarrow E(\tilde{X}) = x_1 p(x_1) + x_2 p(x_2) + x_3 p(x_3)$$

yielding

$$E(\tilde{X}) = x_1 p_1 + x_2 p_2 + x_3 p_3. \tag{1.10}$$

Using (1.9), $E(\tilde{X}^2)$ can be calculated as

$$E(\tilde{X}^2) = \sum_x x^2 p(x) \rightarrow E(\tilde{X}^2) = x_1^2 p(x_1) + x_2^2 p(x_2) + x_3^2 p(x_3)$$

yielding

$$E(\tilde{X}^2) = x_1^2 p_1 + x_2^2 p_2 + x_3^2 p_3. \tag{1.11}$$

And finally, $Var(\tilde{X})$ is calculated using (1.10) and (1.11) as

$$Var(\tilde{X}) = E(\tilde{X}^2) - [E(\tilde{X})]^2$$

yielding

$$Var(\tilde{X}) = x_1^2 p_1 + x_2^2 p_2 + x_3^2 p_3 - [x_1 p_1 + x_2 p_2 + x_3 p_3]^2$$

Example 1.2 For the discrete R.V. $\tilde{X}$, the range set is $R_{\tilde{X}} = \{-1.2, 2.5, 3.2\}$, and the probability mass function for the elements of the range set is defined as

$$p(-1.2) = 1/4 \qquad p(2.5) = 2/4 \qquad p(3.2) = 1/4.$$

Find $E(\tilde{X})$, $E(\tilde{X}^2)$ and $Var(\tilde{X})$.

Solution 1.2 Following the same steps as in the previous example, we can calculate the mean and variance of the given discrete random variable. The mean is calculated as

$$E(\tilde{X}) = \sum_x xp(x) \rightarrow E(\tilde{X}) = -1.2 \times \underbrace{p(-1.2)}_{1/4} + 2.5 \times \underbrace{p(2.5)}_{1/2} + 3.2 \times \underbrace{p(3.2)}_{1/4}$$

yielding

$$E(\tilde{X}) = 1.75.$$

$E(\tilde{X}^2)$ is calculated as

$$E(\tilde{X}^2) = \sum_x x^2 p(x) \rightarrow E(\tilde{X}^2)$$
$$= (-1.2)^2 \times \underbrace{p(-1.2)}_{1/4} + 2.5^2 \times \underbrace{p(2.5)}_{1/2} + 3.2^2 \times \underbrace{p(3.2)}_{1/4}$$

yielding

$$E(\tilde{X}^2) = 6.0450.$$

Finally variance of the discrete random variable $\tilde{X}$ is calculated as

$$Var(\tilde{X}) = E(\tilde{X}^2) - \left[E(\tilde{X})\right]^2 \rightarrow Var(\tilde{X}) = 6.0450 - 1.75^2 \rightarrow Var(\tilde{X}) = 2.9825.$$

1.3 Discrete Entropy

Let $\tilde{X}$ be a discrete random variable, and let the range set of this random variable be given as

$$R_{\tilde{X}} = \{x_1, x_2, \ldots, x_N\}$$

then the information content of the symbol $x_i \in R_{\tilde{X}}$ is

$$I(x_i) = \log\left(\frac{1}{p(x_i)}\right) \rightarrow I(x_i) = -\log(p(x_i)). \quad (1.12)$$

The average information content of the discrete random variable $\tilde{X}$ is defined as

$$H(\tilde{X}) = \sum_{x_i} p(x_i) I(x_i) \quad (1.13)$$

which is nothing but expected value of $I(\tilde{X})$, i.e., $E(I(\tilde{X}))$.

Note:

$$E(g(\tilde{X})) = \sum_{x_i} p(x_i) g(x_i) \quad (1.14)$$

The equation in (1.13) can also be written as

$$H(\tilde{X}) = \sum_{x_i} p(x_i) \log\left(\frac{1}{p(x_i)}\right) \quad (1.15)$$

which is equal to

$$H(\tilde{X}) = -\sum_{x_i} p(x_i) \log(p(x_i)). \quad (1.16)$$

The entropy expression in (1.16) can also be written in a more compact form as

$$H(\tilde{X}) = -\sum_{x} p(x)\log(p(x)) \tag{1.17}$$

where x are the values generated by random variable $\tilde{X}$. The mathematical expression (1.17) can also be expressed as

$$H(\tilde{X}) = -E\left(\log\left(p\left(\tilde{X}\right)\right)\right)$$

where $E(\cdot)$ is the expected value operator for discrete random variable $\tilde{X}$.

Example 1.3 For the discrete R.V. $\tilde{X}$, the range set is $R_{\tilde{X}} = \{x_1, x_2\}$, and the probability mass function for the elements of the range set is defined as

$$p(x_1) = 1/3 \qquad p(x_2) = 2/3$$

Find the information content of each symbol, and find the average information content of the discrete random variable.

Solution 1.3 Using (1.12), the information content of each symbol is found as

$$I(x_1) = \log\left(\frac{1}{p(x_1)}\right) \rightarrow I(x_1) = \log\left(\frac{1}{\frac{1}{3}}\right) \rightarrow I(x_1) = \log(3)$$

$$I(x_2) = \log\left(\frac{1}{p(x_2)}\right) \rightarrow I(x_2) = \log\left(\frac{1}{\frac{2}{3}}\right) \rightarrow I(x_2) = \log(3/2).$$

Note: The base of the logarithm function, i.e., $\log(\cdot)$, is always '2' unless otherwise indicated.

The average information content of the random variable is calculated using (1.15) or (1.16) as

$$H(\tilde{X}) = \sum_{x_i} p(x_i)\log\left(\frac{1}{p(x_i)}\right) \rightarrow H(\tilde{X}) = p(x_1)\log\left(\frac{1}{p(x_1)}\right) + p(x_2)\log\left(\frac{1}{p(x_2)}\right)$$

in which substituting the numerical values, we obtain

$$H(\tilde{X}) = \frac{1}{3}\log(3) + \frac{2}{3}\log\left(\frac{3}{2}\right) \rightarrow H(\tilde{X}) = 0.9183 \text{ bits/symbol}.$$

Example 1.4 For the discrete R.V. $\tilde{X}$, the range set is $R_{\tilde{X}} = \{x_1, x_2\}$, and the probability mass function for the elements of the range set is defined as

$$p(x_1) = p \qquad p(x_2) = 1 - p$$

Find the average information content of the discrete random variable, i.e., find the entropy of the discrete random variable.

Solution 1.4 Using (1.16), the entropy of the discrete random variable can be calculated as

$$H(\tilde{X}) = -\sum_{x_i} p(x_i)\log(p(x_i)) \rightarrow H(\tilde{X}) = -[p(x_1)\log(p(x_1)) + p(x_2)\log(p(x_2))]$$

which is written as

$$H(\tilde{X}) = -[p\log(p) + (1-p)\log(1-p)]. \tag{1.18}$$

The right hand side of (1.18) is a function of p only, and we can express the right hand side of (1.18) by a function $H_b(p)$, i.e.,

$$H_b(p) = -[p\log(p) + (1-p)\log(1-p)]. \tag{1.19}$$

Example 1.5 Plot the graph of the function

$$H_b(p) = -[p\log(p) + (1-p)\log(1-p)]$$

w.r.t. p bounded as $0 \leq p \leq 1$.

Solution 1.5 The graph of $H_b(p) = -[p\log(p) + (1-p)\log(1-p)]$ is plotted using matlab in Fig. 1.1.
As it is clear from Fig. 1.1 that the function $H_b(p) = -[p\log(p) + (1-p)\log(1-p)]$ takes its maximum value at point $p = 0.5$.
We can mathematically calculate the value of p at which $H_b(p)$ gets its maximum by taking the derivative of $H_b(p)$ w.r.t. p and equating it to zero as follows

$$\frac{\partial H_b(p)}{\partial p} = 0 \rightarrow \log p + \ln 2 - \log(1-p) - \ln 2 = 0 \rightarrow p = 0.5.$$

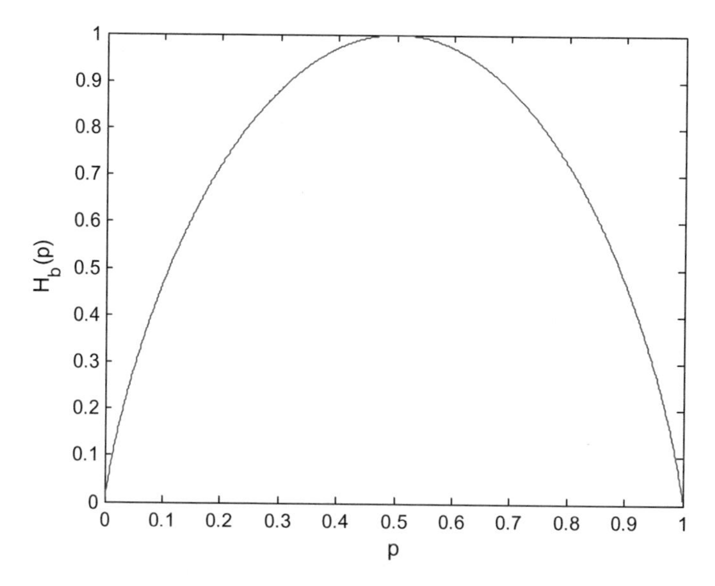

Fig. 1.1 The graph of $H_b(p)$

1.3.1 Interpretation of Entropy

Entropy is the average information content of an information source, i.e., average information content of a random variable. Consider that we are receiving symbols from an information source, if the entropy of the information source is high, it means that we are receiving significant amount of information from the source. Otherwise, the amount of information received from the information source is not large.

Entropy can also be interpreted as the chaos amount of a source. And the average information required to detail the chaotic environment is nothing but the entropy. For instance, consider a country where instability is available, and too many events which cannot be controlled by government forces occur. Assume that we are in a different country and want to learn the situation in that chaotic country. In this case we need too much information to identify the uncertain cases in the problematic country. Hence, we can say that entropy of the problematic country is high.

1.3.2 Joint Entropy

Let $\tilde{X}$ and $\tilde{Y}$ be two discrete random variables with marginal and joint probability mass functions $p_{\tilde{X}}(x)$, $p_{\tilde{Y}}(y)$ and $p_{\tilde{X},\tilde{Y}}(x, y)$ respectively.

Note: $p_{\tilde{X}}(x) = Prob(\tilde{X} = x)$, $p_{\tilde{Y}}(y) = Prob(\tilde{Y} = y)$ and $p_{\tilde{X},\tilde{Y}}(x, y) = Prob(\tilde{X} = x, \tilde{Y} = y)$

For the easy of notation, we will use $p(x)$, $p(y)$, and $p(x, y)$ instead of $p_{\tilde{X}}(x)$, $p_{\tilde{Y}}(y)$, and $p_{\tilde{X},\tilde{Y}}(x, y)$ respectively, from now on throughout the book unless otherwise indicated.

The joint entropy for the discrete random variables $\tilde{X}$ and $\tilde{Y}$ is defined as

$$H(\tilde{X}, \tilde{Y}) = \sum_{x_i, y_j} p(x_i, y_j) \log\left(\frac{1}{p(x_i, y_j)}\right) \tag{1.20}$$

which can be written in a more compact form as

$$H(\tilde{X}, \tilde{Y}) = -\sum_{x,y} p(x, y) \log(p(x, y)) \tag{1.21}$$

Example 1.6 The discrete random variables $\tilde{X}$ and $\tilde{Y}$ have the range sets $R_{\tilde{X}} = \{x_1, x_2\}$, and $R_{\tilde{Y}} = \{y_1, y_2, y_3\}$. Find $H(\tilde{X}, \tilde{Y})$.

Solution 1.6 Expanding

$$H(\tilde{X}, \tilde{Y}) = -\sum_{x_i, y_j} p(x_i, y_j) \log(p(x_i, y_j))$$

for x_1, x_2 we obtain

$$H(\tilde{X}, \tilde{Y}) = -\sum_{x_1, y_j} p(x_1, y_j) \log(p(x_1, y_j)) - \sum_{x_2, y_j} p(x_2, y_j) \log(p(x_2, y_j)). \tag{1.22}$$

In the next step, expanding the right hand side of (1.22) for y_1, y_2, y_3, we get

$$\begin{aligned} H(\tilde{X}, \tilde{Y}) = &-[p(x_1, y_1)\log(p(x_1, y_1)) + p(x_1, y_2)\log(p(x_1, y_2)) + p(x_1, y_3)\log(p(x_1, y_3))] + \\ &- p(x_2, y_1)\log(p(x_2, y_1)) + p(x_2, y_2)\log(p(x_2, y_2)) + p(x_2, y_3)\log(p(x_2, y_3))) \end{aligned}$$

Remark Let $|R_{\tilde{X}}|$ and $|R_{\tilde{Y}}|$ be the number of elements in the range sets of $\tilde{X}$ and $\tilde{Y}$ respectively. The number of $p(x, y)\log(p(x, y))$ terms in $H(\tilde{X}, \tilde{Y})$ expression equals to $|R_{\tilde{X}}| \times |R_{\tilde{Y}}|$.

1.3.3 Conditional Entropy

The conditional entropy of the discrete random variable $\tilde{X}$ for a given value y_j of another discrete random variable $\tilde{Y}$ is defined as

$$H(\tilde{X}|y_j) = -\sum_{x_i} p(x_i|y_j)\log(p(x_i|y_j)). \tag{1.23}$$

which can be written in a more compact form as

$$H(\tilde{X}|y) = -\sum_{x} p(x|y)\log(p(x|y)). \tag{1.24}$$

Equation (1.23) can be interpreted as the amount of average information provided by a single symbol of random variable $\tilde{X}$ if a single symbol of random variable $\tilde{Y}$ is known, or it can be interpreted in a different way as the amount of average information required to know a single symbol of random variable $\tilde{X}$ if a single symbol of random variable $\tilde{Y}$ is known.

If $\tilde{X}$ and $\tilde{Y}$ are not independent random variables, it is obvious that $H(\tilde{X}|y) < H(\tilde{X})$. This means that we need less amount of average information per symbol to know the random variable $\tilde{X}$, since a value of $\tilde{Y}$ provides some information about the random variable $\tilde{X}$, or the random variable $\tilde{X}$ provides us with less amount of average information per symbol, since a value of $\tilde{Y}$ also provided some information.

The conditional entropy of the discrete random variable $\tilde{X}$ given another discrete random variable $\tilde{Y}$, i.e., all the values (symbols) of $\tilde{Y}$ are known, is defined as

$$H(\tilde{X}|\tilde{Y}) = \sum_{y_j} p(y_j)H(\tilde{X}|y_j) \tag{1.25}$$

which can also be written in a more general way as

$$H(\tilde{X}|\tilde{Y}) = \sum_{y} p(y)H(\tilde{X}|y). \tag{1.26}$$

Equation (1.25) can be considered as the total amount of average information per symbol required to know the random variable $\tilde{X}$ assuming that the random variable $\tilde{Y}$ is known, i.e., all the values (symbols) of $\tilde{Y}$ are known, and each symbol provides some information about $\tilde{X}$, and we consider average amount of information per symbol to know the random variable $\tilde{X}$ omitting the amount of information provided by the symbols of $\tilde{Y}$ accounting their probability of occurrence.

Equation (1.25) can also be considered as the total amount of information provided by the random variable $\tilde{X}$ assuming that the random variable $\tilde{Y}$ is known, i.e., all the values (symbols) of $\tilde{Y}$ are known and each symbol provides some information available in $\tilde{X}$, and we consider total amount of information provided by $\tilde{X}$ excluding the amount of information provided by the symbols of $\tilde{Y}$ accounting their probability of occurrence.

Substituting (1.23) into (1.25), we obtain

$$H(\tilde{X}|\tilde{Y}) = -\sum_{y_j} p(y_j) \sum_{x_i} p(x_i|y_j) \log(p(x_i|y_j))$$

which can be written as

$$H(\tilde{X}|\tilde{Y}) = -\sum_{x_i, y_j} p(x_i|y_j) p(y_j) \log(p(x_i|y_j))$$

where employing $p(x_i|y_j)p(y_j) = p(x_i, y_j)$, we obtain

$$H(\tilde{X}|\tilde{Y}) = -\sum_{x_i, y_j} p(x_i, y_j) \log(p(x_i|y_j)) \tag{1.27}$$

which can be written in a more compact form as

$$H(\tilde{X}|\tilde{Y}) = -\sum_{x,y} p(x, y) \log(p(x|y)). \tag{1.28}$$

A more general expression of conditional entropy considering n different discrete random variables can be defined as

$$H(\tilde{X}|\tilde{Y}_1, \tilde{Y}_2, \ldots, \tilde{Y}_{n-1}) = -\sum_{x, y_1, y_2, \ldots y_{n-1}} p(x, y_1, y_2, \ldots y_{n-1}) \log(p(x|y_1, y_2, \ldots y_{n-1}))$$

where

$$p(x, y_1, y_2, \ldots, y_{n-1}) = Prob(\tilde{X} = x, \tilde{Y}_1 = y_1, \tilde{Y}_2 = y_2, \ldots, \tilde{Y}_{n-1} = y_{n-1}).$$

Example 1.7 The discrete random variables $\tilde{X}$ and $\tilde{Y}$ have the range sets $R_{\tilde{X}} = \{x_1, x_2, x_3\}$, and $R_{\tilde{Y}} = \{y_1, y_2\}$. Find $H(\tilde{X}|y)$.

Solution 1.7 In $H(\tilde{X}|y)$ conditional entropy, y is a general parameter for the values of $\tilde{Y}$. Since, the range set of $\tilde{Y}$ is $R_{\tilde{Y}} = \{y_1, y_2\}$, then y can be either equal to y_1 or equal to y_2. For this reason, we need to calculate $H(\tilde{X}|y_1)$ and $H(\tilde{X}|y_2)$ separately. The calculation of $H(\tilde{X}|y_1)$ can be achieved using

$$H(\tilde{X}|y_1) = -\sum_{x} p(x|y_1)\log(p(x|y_1)). \tag{1.29}$$

When (1.29) is expanded, we obtain

$$H(\tilde{X}|y_1) = -[p(x_1|y_1)\log(p(x_1|y_1)) + p(x_2|y_1)\log(p(x_2|y_1)) + p(x_3|y_1)\log(p(x_3|y_1))].$$

In a similar manner, $H(\tilde{X}|y_2)$ can be calculated via

$$H(\tilde{X}|y_2) = -\sum_{x} p(x|y_2)\log(p(x|y_2)). \tag{1.30}$$

When (1.30) is expanded, we obtain

$$H(\tilde{X}|y_2) = -[p(x_1|y_2)\log(p(x_1|y_2)) + p(x_2|y_2)\log(p(x_2|y_2)) + p(x_3|y_2)\log(p(x_3|y_2))].$$

Example 1.8 The discrete random variables $\tilde{X}$ and $\tilde{Y}$ have the range sets $R_{\tilde{X}} = \{x_1, x_2, x_3\}$, and $R_{\tilde{Y}} = \{y_1, y_2\}$. Find $H(\tilde{X}|\tilde{Y})$.

Solution 1.8 The conditional entropy $H(\tilde{X}|\tilde{Y})$ can be calculated using either

$$H(\tilde{X}|\tilde{Y}) = \sum_{y_j} p(y_j)H(\tilde{X}|y_j) \tag{1.31}$$

or using

$$H(\tilde{X}|\tilde{Y}) = -\sum_{x_i, y_j} p(x_i, y_j)\log(p(x_i|y_j)). \tag{1.32}$$

Let's use (1.31) to calculate the conditional entropy. Expanding (1.31), we obtain

$$H(\tilde{X}|\tilde{Y}) = p(y_1)H(\tilde{X}|y_1) + p(y_2)H(\tilde{X}|y_2). \tag{1.33}$$

The expressions $H(\tilde{X}|y_1)$ and $H(\tilde{X}|y_2)$ in (1.33) are calculated as

$$H(\tilde{X}|y_1) = -[p(x_1|y_1)\log(p(x_1|y_1)) + p(x_2|y_1)\log(p(x_2|y_1)) + p(x_3|y_1)\log(p(x_3|y_1))]$$

$$H(\tilde{X}|y_2) = -[p(x_1|y_2)\log(p(x_1|y_2)) + p(x_2|y_2)\log(p(x_2|y_2)) + p(x_3|y_2)\log(p(x_3|y_2))].$$

Substituting the calculated expressions into (1.33), we obtain

$$\begin{aligned} H(\tilde{X}|\tilde{Y}) = & -[p(x_1,y_1)\log(p(x_1|y_1)) + p(x_2,y_1)\log(p(x_2|y_1)) + p(x_3,y_1)\log(p(x_3|y_1))] \\ & -[p(x_1,y_2)\log(p(x_1|y_2)) + p(x_2,y_2)\log(p(x_2|y_2)) + p(x_3,y_2)\log(p(x_3|y_2))]. \end{aligned}$$

Note: $p(y)p(x|y) = p(x,y)$

Example 1.9 Consider the discrete memoryless communication channel in Fig. 1.2. Find $H(\tilde{X}|\tilde{Y})$.

Solution 1.9 The conditional entropy $H(\tilde{X}|\tilde{Y})$ can be interpreted as the amount of remaining uncertainty about $\tilde{X}$ when $\tilde{Y}$ is known, in other words, the amount of average information per-symbol needed to know the random variable $\tilde{X}$ when the random variable $\tilde{Y}$ is known. It is clear from the communication channel that when $\tilde{Y}$ is known, $\tilde{X}$ is also known. That is, assume that y_1 is received, then we can definitely know the transmitted symbol, i.e., we can decide that x_1 is transmitted. In that case, no additional information is required to identify the transmitted symbol. Hence, we can conclude that $H(\tilde{X}|\tilde{Y}) = 0$.

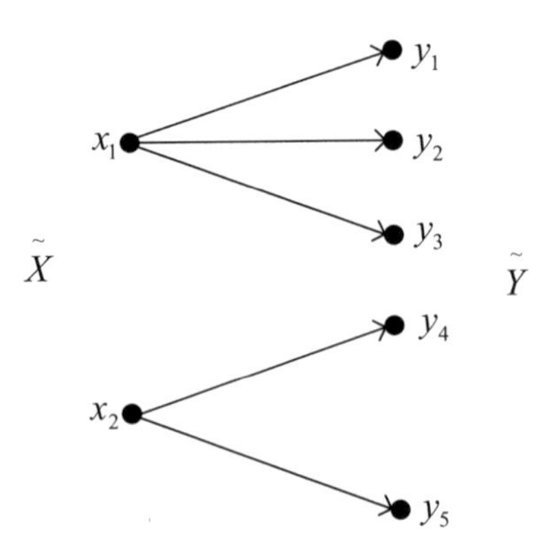

Fig. 1.2 Discrete communication channel for Example 1.9

Let's now mathematically prove that $H(\tilde{X}|\tilde{Y}) = 0$. For this purpose, let's use the formula

$$H(\tilde{X}|\tilde{Y}) = \sum_{y_j} p(y_j) H(\tilde{X}|y_j) \tag{1.34}$$

where

$$H(\tilde{X}|y_j) = -\sum_{x_i} p(x_i|y_j) \log(p(x_i|y_j)). \tag{1.35}$$

It is obvious from (1.35) that we need to calculate the conditional probabilities $p(x_i|y_j)$ between channel inputs and outputs. We can write $p(x_i|y_j)$ as

$$p(x_i|y_j) = \frac{p(y_j|x_i)p(x_i)}{p(y_j)} \tag{1.36}$$

where $p(y_j|x_i)$ is the probability of receiving y_j when x_i is transmitted. If there is no connection between a channel input x_i and a channel output y_j, then it is obvious that

$$p(y_j|x_i) = 0.$$

And from (1.36), we can also conclude that if there is no connection between a channel input x_i and a channel output y_j, then we have

$$p(x_i|y_j) = 0.$$

Considering Fig. 1.2, we can write the following probabilities

$$p(x_1|y_4) = 0, \quad p(x_1|y_5) = 0$$

$$p(x_2|y_1) = 0, \quad p(x_2|y_2) = 0, \quad p(x_2|y_3) = 0.$$

In addition, we know that

$$0 \log(0) = 0.$$

Hence, we can write that

$$p(x_1|y_4) \log(p(x_1|y_4)) = 0, \qquad p(x_1|y_5) \log(p(x_1|y_5)) = 0,$$

$$p(x_2|y_1) \log(p(x_2|y_1)) = 0, \qquad p(x_2|y_2) \log(p(x_2|y_2)) = 0,$$

$$p(x_2|y_3)\log(p(x_2|y_3)) = 0.$$

If there is direct connection between a channel input x_i and a channel output y_j, then $p(y_j)$ can be calculated as

$$p(y_j) = p(y_j|x_i)p(x_i). \tag{1.37}$$

Note:

$$p(y) = \sum_x p(x,y) \rightarrow p(y) = \sum_x p(y|x)p(x)$$

The probability expression $p(x_i|y_j)$, between directly connected input x_i and output y_j, can be calculated using

$$p(x_i|y_j) = \frac{p(y_j|x_i)p(x_i)}{p(y_j)}$$

where substituting (1.37) for $p(y_j)$, we get

$$p(x_i|y_j) = \frac{p(y_j|x_i)p(x_i)}{p(y_j|x_i)p(x_i)} \rightarrow p(x_i|y_j) = 1.$$

This means that for the directly connected input x_i and output y_j, we have

$$p(x_i|y_j)\log(p(x_i|y_j)) = 0.$$

Considering the direct connected inputs and outputs, we can write

$$p(x_1|y_1)\log(p(x_1|y_1)) = 0, \qquad p(x_1|y_2)\log(p(x_1|y_2)) = 0,$$

$$p(x_1|y_3)\log(p(x_1|y_3)) = 0,$$

$$p(x_2|y_4)\log(p(x_2|y_4)) = 0, \qquad p(x_2|y_5)\log(p(x_2|y_5)) = 0.$$

Now, if we go to the calculation of

$$H(\tilde{X}|y_j) = -\sum_{x_i} p(x_i|y_j)\log(p(x_i|y_j))$$

for y_1, y_2, y_3, y_4, and y_5 it can be easily verified that $H(\tilde{X}|y_j) = 0, j = 1, \ldots, 5$. For illustration purposes, lets expand the above expression for $H(\tilde{X}|y_1)$ as

$$H(\tilde{X}|y_1) = -\left[\underbrace{p(x_1|y_1)}_{=1}\underbrace{\log(p(x_1|y_1))}_{=0} + \underbrace{p(x_2|y_1)}_{=1}\underbrace{\log(p(x_2|y_1))}_{=0}\right]$$

leading to

$$H(\tilde{X}|y_1) = 0.$$

Since

$$H(\tilde{X}|\tilde{Y}) = \sum_{y_j} p(y_j) H(\tilde{X}|y_j)$$

we can write that

$$H(\tilde{X}|\tilde{Y}) = 0.$$

Example 1.10 For the discrete communication channel shown in Fig. 1.3, the input symbol probabilities and the transition probabilities are given as

$$p(x_1) = \frac{1}{2} \qquad p(x_2) = \frac{1}{2}$$

$$p(y_1|x_1) = \frac{1}{4} \qquad p(y_1|x_2) = \frac{1}{8} \qquad p(y_2|x_1) = \frac{3}{4} \qquad p(y_2|x_2) = \frac{7}{8}.$$

Compare the values of $H(\tilde{X}|y_1)$ and $H(\tilde{X}|y_2)$ to each other without mathematically calculating them.

Solution 1.10 Let's consider $H(\tilde{X}|y_1)$ and $H(\tilde{X}|y_2)$ separately. $H(\tilde{X}|y_1)$ is the amount of uncertainty about $\tilde{X}$ when y_1 is known. The transmission of x_1, x_2 and the reception of y_1 and y_2 are separately shown in Fig. 1.3.

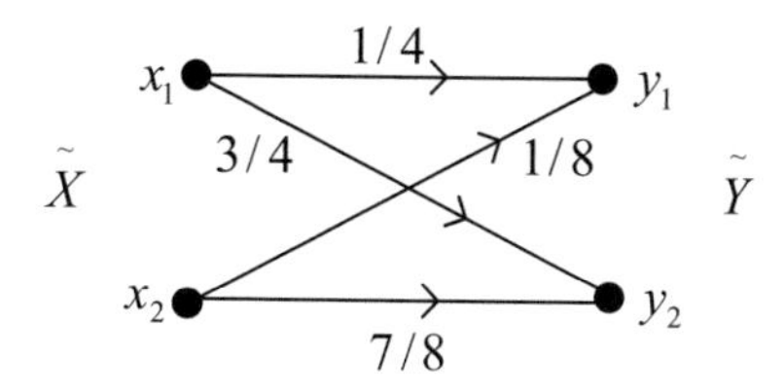

Fig. 1.3 Discrete communication channel for Example 1.10

If take the ratio of channel transition probabilities in Fig. 1.4a, we obtain $\frac{1/4}{1/8}=2$ or $\frac{\frac{1}{8}}{\frac{1}{4}}=0.5$, and similarly, if we take the ratio of channel transition probabilities in Fig. 1.4b, we obtain $\frac{3/4}{7/8}=0.86$ or $\frac{\frac{7}{8}}{\frac{3}{4}}=1.16$. This means that the transition probabilities in Fig. 1.4a are far away from each other, on the other hand, the channel transition probabilities in Fig. 1.4b are close to each other. This means that if y_1 is received, we can estimate the transmitted symbol with a larger probability, in other words, uncertainty about the identity of the transmitted symbol identity is less when the system in Fig. 1.4a is considered. When y_2 is received, we cannot make an estimate as accurate as for the case when y_1 is received. Since, transition probabilities of both symbols are high and close to each other for the system in Fig. 1.4b. This means that when y_2 is received, the uncertainty about $\tilde{X}$ is high. It also implies that conditional entropy of $\tilde{X}$ is high. Hence, we can write that

$$H(\tilde{X}|y_2) > H(\tilde{X}|y_1).$$

Don't forget that the entropy is related to the uncertainty amount of the source. If uncertainty is high, then entropy is also high.

Example 1.11 For the previous example, without mathematically calculating $H(\tilde{Y}|x_1)$ and $H(\tilde{Y}|x_2)$, decide which one is greater than the other.

Solution 1.11 Following a similar reasoning as in the previous example we can find that

$$H(\tilde{Y}|x_1) > H(\tilde{Y}|x_2).$$

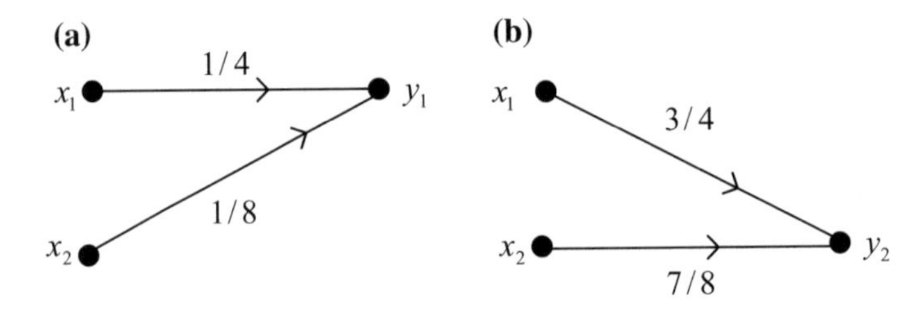

Fig. 1.4 Separation of the channel

Example 1.12 For the discrete communication channel shown in Fig. 1.5, the input symbol probabilities and the transition probabilities are given as

$$p(x_1) = \frac{1}{2} \qquad p(x_2) = \frac{1}{2}$$

$$p(y_1|x_1) = \frac{1}{8} \qquad p(y_1|x_2) = \frac{1}{4} \qquad p(y_2|x_1) = \frac{2}{8} \qquad p(y_2|x_2) = \frac{1}{4}$$

$$p(y_3|x_1) = \frac{5}{8} \qquad p(y_3|x_2) = \frac{2}{4}.$$

Compare the values of $H(\tilde{X}|y_1)$, $H(\tilde{X}|y_2)$, and $H(\tilde{X}|y_3)$ without mathematically calculating them.

Solution 1.12 The discrete memoryless communication channel show in Fig. 1.5 can be decomposed for each output symbol as shown in Fig. 1.6 where channel transition probabilities $p(y|x)$ are shown separately for each output symbol. For Fig. 1.6a, the ratio of the transition probabilities is

$$\frac{\frac{1}{8}}{\frac{1}{4}} = \frac{1}{2}.$$

In a similar manner for Fig. 1.6b, c the ratios of the channel transition probabilities are

$$\frac{\frac{2}{8}}{\frac{1}{4}} = 1 \qquad \frac{\frac{5}{8}}{\frac{2}{4}} = \frac{5}{4}.$$

And we know that as the ratio approaches to 1, it becomes more difficult to estimate the transmitted symbol considering the received symbol. Since the likelihood of

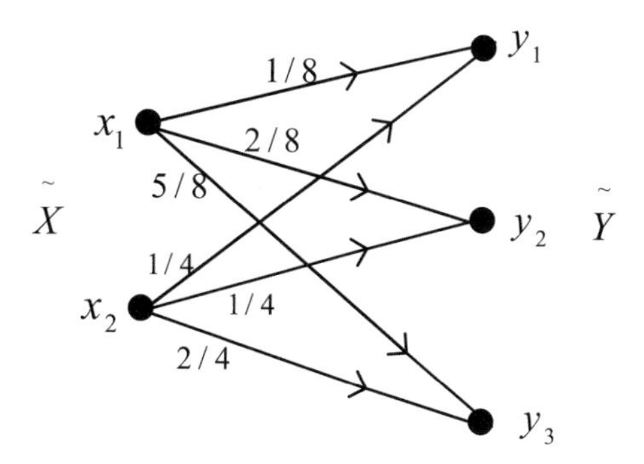

Fig. 1.5 Discrete communication channel for Example 1.12

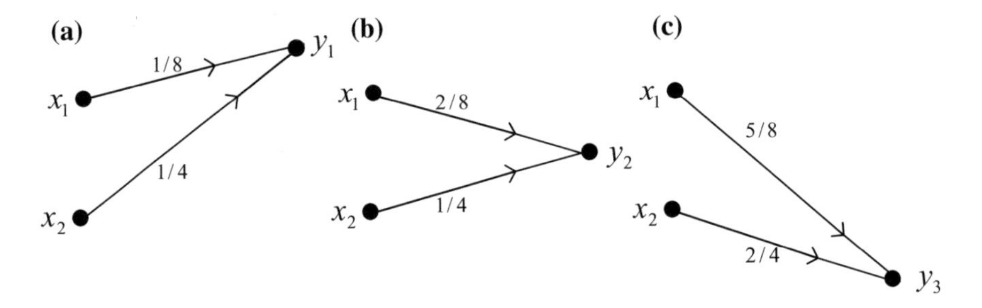

Fig. 1.6 Separation of the channel

transmitting each symbol approaches to each other. This means that the uncertainty of the source is high, i.e., entropy of the source is high.

On the other hand, as the ratio of the transition probabilities gets far away from '1', then the transmitted source symbol can be estimated with a higher probability. This also means that the uncertainty of the source is low, i.e., entropy of the source is low. Considering these two issues, we can order the conditional entropies as

$$H\left(\tilde{X}|y_2\right) > H\left(\tilde{X}|y_3\right) > H\left(\tilde{X}|y_1\right).$$

Exercise For the previous example, without mathematically calculating $H\left(\tilde{Y}|x_1\right)$ and $H\left(\tilde{Y}|x_2\right)$, decide which one is greater than the other.

Example 1.13 For the discrete communication channel shown in Fig. 1.7, the input symbol probabilities, and transition probabilities are given as

$$p(x_1) = \frac{1}{2} \qquad p(x_2) = \frac{1}{2}$$

$$p(y_1|x_1) = \frac{1}{4} \qquad p(y_1|x_2) = \frac{1}{8} \qquad p(y_2|x_1) = \frac{3}{4} \qquad p(y_2|x_2) = \frac{7}{8}.$$

Calculate the following

$$p(y_1) \qquad p(y_2) \qquad H\left(\tilde{X}|y_1\right) \qquad H\left(\tilde{X}|y_2\right) \qquad H\left(\tilde{Y}|x_1\right) \qquad H\left(\tilde{Y}|x_2\right).$$

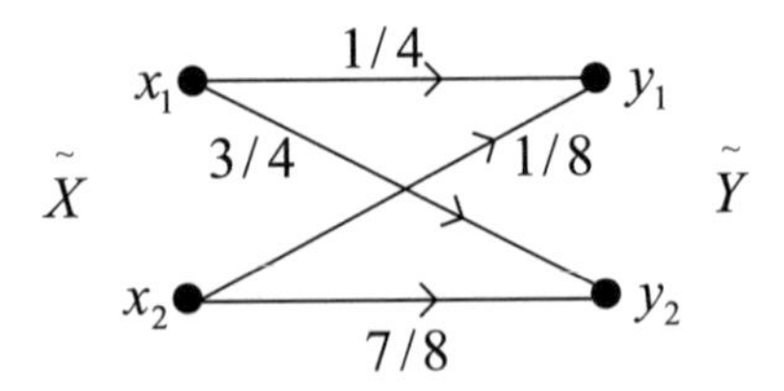

Fig. 1.7 Discrete communication channel for Example 1.13

Solution 1.13 It is known that the marginal probability density function, i.e., pdf, $p(y)$ can be calculated from joint pdf $p(x,y)$ using

$$p(y)=\sum_x p(x,y)\rightarrow p(y)=\sum_x p(y|x)p(x).$$

Then, for $p(y_1)$ we have

$$p(y_1)=\sum_x p(y_1|x)p(x)\rightarrow p(y_1)=\underbrace{p(y_1|x_1)}_{1/4}\underbrace{p(x_1)}_{1/2} + \underbrace{p(y_1|x_2)}_{1/8}\underbrace{p(x_2)}_{1/2}\rightarrow p(y_1)=3/16.$$

In a similar manner, for $p(y_2)$ we have

$$p(y_2)=\sum_x p(y_2|x)p(x)\rightarrow p(y_2)=\underbrace{p(y_2|x_1)}_{3/4}\underbrace{p(x_1)}_{1/2} + \underbrace{p(y_2|x_2)}_{7/8}\underbrace{p(x_2)}_{1/2}\rightarrow p(y_2)=13/16.$$

Remark $p(y_1)+p(y_2)=1$

For the calculation of $H(\tilde{X}|y_1)$, we can use

$$H(\tilde{X}|y_1)=-\sum_x p(x|y_1)\log(p(x|y_1)). \tag{1.38}$$

When (1.38) is expanded, we obtain

$$H(\tilde{X}|y_1)=-[p(x_1|y_1)\log(p(x_1|y_1))+p(x_2|y_1)\log(p(x_2|y_1))] \tag{1.39}$$

where the conditional probabilities $p(x_1|y_1)$ and $p(x_2|y_1)$ can be calculated as

$$\begin{aligned} p(x_1|y_1)&=\frac{p(y_1|x_1)p(x_1)}{p(y_1)}\rightarrow p(x_1|y_1)=\frac{\frac{1}{4}\times\frac{1}{2}}{\frac{3}{16}}\rightarrow p(x_1|y_1)=\frac{2}{3}\\ p(x_2|y_1)&=\frac{p(y_1|x_2)p(x_2)}{p(y_1)}\rightarrow p(x_2|y_1)=\frac{\frac{1}{8}\times\frac{1}{2}}{\frac{3}{16}}\rightarrow p(x_2|y_1)=\frac{1}{3}. \end{aligned} \tag{1.40}$$

Remark

$$\sum_x p(x|y)=1$$

Substituting the calculated probabilities in (1.40) into (1.39), we obtain

$$H(\tilde{X}|y_1) = \left[\frac{2}{3}\log\left(\frac{2}{3}\right) + \frac{1}{3}\log\left(\frac{1}{3}\right)\right] \rightarrow H(\tilde{X}|y_1) = 0.9183 \text{ bits/symbol.}$$

For the calculation of $H(\tilde{X}|y_2)$, we can use

$$H(\tilde{X}|y_2) = -\sum_x p(x|y_2)\log(p(x|y_2)). \tag{1.41}$$

When (1.41) is expanded, we obtain

$$H(\tilde{X}|y_2) = -[p(x_1|y_2)\log(p(x_1|y_2)) + p(x_2|y_2)\log(p(x_2|y_2))] \tag{1.42}$$

where the conditional probabilities $p(x_1|y_2)$ and $p(x_2|y_2)$ can be calculated

$$\begin{aligned} p(x_1|y_2) &= \frac{p(y_2|x_1)p(x_1)}{p(y_2)} \rightarrow p(x_1|y_2) = \frac{\frac{3}{4}\times\frac{1}{2}}{\frac{13}{16}} \rightarrow p(x_1|y_2) = \frac{6}{13} \\ p(x_2|y_2) &= \frac{p(y_2|x_2)p(x_2)}{p(y_2)} \rightarrow p(x_2|y_2) = \frac{\frac{7}{8}\times\frac{1}{2}}{\frac{13}{16}} \rightarrow p(x_2|y_2) = \frac{7}{13}. \end{aligned} \tag{1.43}$$

Substituting the calculated probabilities in (1.43) into (1.42), we obtain

$$H(\tilde{X}|y_2) = -\left[\frac{6}{13}\log\left(\frac{6}{13}\right) + \frac{7}{13}\log\left(\frac{7}{13}\right)\right] \rightarrow H(\tilde{X}|y_2) = 0.9957 \text{ bits/symbol.}$$

For the calculation of $H(\tilde{Y}|x_1)$, we can use

$$H(\tilde{Y}|x_1) = -\sum_y p(y|x_1)\log(p(y|x_1)). \tag{1.44}$$

When (1.44) is expanded, we obtain

$$H(\tilde{Y}|x_1) = -[p(y_1|x_1)\log(p(y_1|x_1)) + p(y_2|x_1)\log(p(y_2|x_1))]. \tag{1.45}$$

Substituting the channel transition probabilities given in the question into (1.45), we obtain

$$H(\tilde{Y}|x_1) = -\left[\frac{1}{4}\log\left(\frac{1}{4}\right) + \frac{3}{4}\log\left(\frac{3}{4}\right)\right] \rightarrow H(\tilde{Y}|x_1) = 0.8113 \text{ bits/symbol.}$$

Following similar steps, we can calculate $H(\tilde{Y}|x_2)$ as

$$H(\tilde{Y}|x_2) = -\left[\frac{1}{8}\log\left(\frac{1}{8}\right) + \frac{7}{8}\log\left(\frac{7}{8}\right)\right] \rightarrow H(\tilde{Y}|x_1) = 0.5436\,\text{bits/symbol}.$$

Although it is not asked in the question, besides, we can compute $H(\tilde{X}|\tilde{Y})$ or $H(\tilde{Y}|\tilde{X})$ using the found results.

Exercise For the discrete communication channel shown in Fig. 1.8, the input symbol probabilities and transition probabilities are given as

$$p(x_1) = \frac{1}{8} \qquad p(x_2) = \frac{7}{8}$$

$$p(y_1|x_1) = \frac{1}{4} \qquad p(y_1|x_2) = \frac{1}{8} \qquad p(y_2|x_1) = \frac{3}{4} \qquad p(y_2|x_2) = \frac{7}{8}.$$

Find output symbol probabilities, and find $H(\tilde{X})$, $H(\tilde{Y})$ and comment on them.

Example 1.14 For a discrete memoryless source, the entropy is given as $H(\tilde{X}) = 4\,\text{bits/sym}$. If we receive 20 symbols from the source approximately, how much information in total, we receive from the source?

Solution 1.14 Entropy is the average information content of an information source. For a sequence consisting of 20 symbols, it is not possible to calculate the amount of information supplied if the symbols forming the sequence are not known.

1.3.4 Properties of the Discrete Entropy

If $\tilde{X}$ and $\tilde{Y}$ are two discrete random variables, and $H(\tilde{X})$ and $H(\tilde{Y})$ are the corresponding entropies, then we have the following properties:

(1) Discrete entropy is a non-negative quantity, i.e., $H(\tilde{X}) \geq 0$.
(2) If the number of elements in the range set of $\tilde{X}$ equals to $|R_{\tilde{X}}|$, the maximum entropy of $\tilde{X}$ equals to

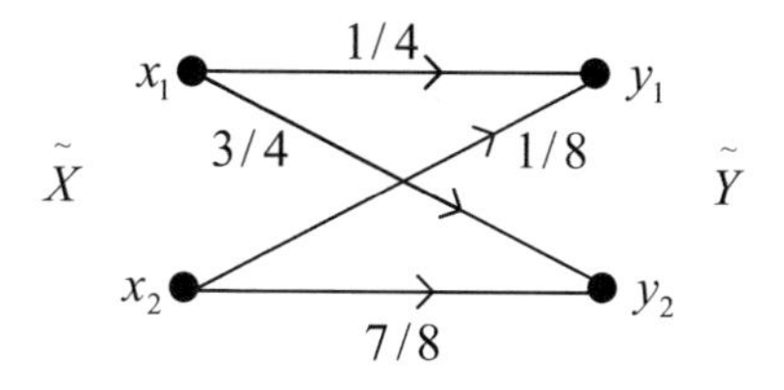

Fig. 1.8 Discrete communication channel for exercise

$$H_{max}(\tilde{X}) = \log(|R_{\tilde{X}}|).$$

(3) For the joint entropy $H(\tilde{X}, \tilde{Y})$, we have the property

$$H(\tilde{X}, \tilde{Y}) \leq H(\tilde{X}) + H(\tilde{Y}) \tag{1.46}$$

where equality occurs if $\tilde{X}$ and $\tilde{Y}$ are independent random variables, i.e., $p(x, y = p(x)p(y)$.

(4) For N discrete random variables, property-3 can be generalized as

$$H(\tilde{X}_1, \tilde{X}_2, \ldots, \tilde{X}_N) \leq H(\tilde{X}_1) + H(\tilde{X}_2) + \cdots + H(\tilde{X}_N)$$

where equality occurs if $\tilde{X}_1, \tilde{X}_2, \ldots, \tilde{X}_N$ are independent random variables.

(5) Conditional entropy satisfy

$$H(\tilde{X}|\tilde{Y}) \leq H(\tilde{X}).$$

We can prove all these properties mathematically. However, before starting to the proofs of the properties, let's give some information about log-sum inequality which is a useful inequality employed for the proofs of the properties.

1.3.5 *Log-Sum Inequality*

Lemma *Let*

$$P = \{p_i\} \text{ and } Q = \{q_i\}, \quad i = 1, \ldots, N$$

be two sets of discrete probabilities such that

$$\sum_i p_i = \sum_i q_i = 1$$

then, we have the inequality

$$\sum_i q_i \log q_i \geq \sum_i q_i \log p_i \tag{1.47}$$

where equality occurs if $p_i = q_i$. *The inequality in* (1.47) *is called log-sum inequality in the literature.*

More generally, log-sum inequality can be defined for two distributions $p(x)$ and $q(x)$ as

$$\sum_x p(x) \log p(x) \geq \sum_x p(x) \log q(x) \tag{1.48}$$

or in terms of joint probability mass functions $p(x, y)$ and $q(x, y)$, the log-sum inequality *can be stated as*

$$\sum_{x,y} p(x, y) \log p(x, y) \geq \sum_{x,y} p(x, y) \log q(x, y). \tag{1.49}$$

Joint distributions involving more than two random variables like $p(x, y, z)$ and $q(x, y, z)$ can also be used in log-sum inequality. For three random variable case, the log-sum inequality becomes as

$$\sum_{x,y,z} p(x, y, z) \log p(x, y, z) \geq \sum_{x,y,z} p(x, y, z) \log q(x, y, z). \tag{1.50}$$

Let's now solve an example to illustrate the log-sum inequality.

Example 1.15 Two sets of probabilities are given as

$$P = \left\{ \underbrace{\frac{1}{4}}_{p_1}, \underbrace{\frac{1}{4}}_{p_2}, \underbrace{\frac{1}{2}}_{p_3} \right\} \quad Q = \left\{ \underbrace{\frac{1}{8}}_{q_1}, \underbrace{\frac{4}{8}}_{q_2}, \underbrace{\frac{3}{8}}_{q_3} \right\}.$$

Verify the log-sum inequality in (1.47) using the given probabilities.

Solution 1.15 To verify log-sum inequality, we can either use

$$\sum_i q_i \log q_i \geq \sum_i q_i \log p_i \tag{1.51}$$

or use

$$\sum_i p_i \log p_i \geq \sum_i p_i \log q_i. \tag{1.52}$$

If we expand (1.51), we obtain

$$\underbrace{\frac{1}{8}\log\frac{1}{8}}_{-0.375} + \underbrace{\frac{4}{8}\log\frac{4}{8}}_{-0.5} + \underbrace{\frac{3}{8}\log\frac{3}{8}}_{-0.5306} \geq \underbrace{\frac{1}{8}\log\frac{1}{4}}_{-0.25} + \underbrace{\frac{4}{8}\log\frac{1}{4}}_{-1} + \underbrace{\frac{3}{8}\log\frac{1}{2}}_{-0.3750}. \tag{1.53}$$

When (1.53) is calculated, we get

$$-1.4056 \geq -1.6250 \rightarrow 1.4056 \leq 1.6250\surd$$

which is a correct inequality. In a similar manner, if we expand (1.52), we obtain

$$\underbrace{\frac{1}{4}\log\frac{1}{4}}_{-0.5}+\underbrace{\frac{1}{4}\log\frac{1}{4}}_{-0.5}+\underbrace{\frac{1}{2}\log\frac{1}{2}}_{-0.5} \geq \underbrace{\frac{1}{4}\log\frac{1}{8}}_{-0.75}+\underbrace{\frac{1}{4}\log\frac{4}{8}}_{-0.25}+\underbrace{\frac{1}{2}\log\frac{3}{8}}_{-0.7075}. \quad (1.54)$$

When (1.54) is calculated, we get

$$-1.5 \geq -1.7075 \rightarrow 1.5 \leq 1.7075\surd$$

which is a correct inequality.

Example 1.16 Prove the following property

$$H(\tilde{X}) \geq 0.$$

Solution 1.16 The entropy $H(\tilde{X})$ is calculated using

$$H(\tilde{X}) = \sum_x p(x)\log\frac{1}{p(x)}$$

where $0 \leq p(x) \leq 1$ which implies that

$$\log\frac{1}{p(x)} \geq 0.$$

Thus, we have

$$H(\tilde{X}) \geq 0.$$

Example 1.17 If $q_i = p(x)p(y)$, show that

$$\sum_i q_i = 1.$$

Solution 1.17 In q_i, we can consider the index i as a pair of two real numbers x, and y, i.e., $i = (x, y)$. In other words, two different real numbers are indicated by a single index value, i.e., by i. For instance, if $q_1 = p(x = -0.4)p(y = 1.2)$, then the index of q_1 i.e., '1' implies $x = -0.4$ and $y = 1.2$.

Then, it is obvious that

$$\sum_i q_i = \sum_{x,y} p(x)p(y). \tag{1.55}$$

From (1.55), we get

$$\sum_{x,y} p(x)p(y) = \sum_x \sum_y p(x)p(y) \to \underbrace{\sum_x p(x)}_{=1} \underbrace{\sum_y p(y)}_{=1} = 1$$

Example 1.18 For the discrete random variables $\tilde{X}$ and $\tilde{Y}$, the marginal and joint probability density functions are given as $p(x), p(y)$, and $p(x, y)$. Show that

$$\sum_{x,y} p(x,y) \log p(x,y) \geq \sum_{x,y} p(x,y) \log p(x)p(y). \tag{1.56}$$

Solution 1.18 Let $p_i = p(x, y)$ and $q_i = p(x)p(y)$. Then, it is obvious that

$$\sum_i p_i = \sum_{x,y} p(x,y) \to \sum_i p_i = 1$$

and

$$\sum_i q_i = \sum_{x,y} p(x)p(y) \to \sum_i q_i = 1.$$

Then, the inequality (1.56) given in the question can be written as

$$\sum_i p_i \log p_i \geq \sum_i p_i \log q_i$$

which is nothing but the log-sum inequality, hence, the inequality (1.56) is correct.

Example 1.19 Show that

$$-\sum_{x,y} p(x,y) \log(p(x))$$

equals to

$$-\sum_x p(x) \log(p(x)).$$

Solution 1.19 Expanding the double summation

$$-\sum_{x,y} p(x,y)\log(p(x))$$

we get

$$-\sum_{x}\sum_{y} p(x,y)\log(p(x))$$

which can be written as

$$-\sum_{x}\log(p(x))\underbrace{\sum_{y} p(x,y)}_{p(x)}$$

where the second summation equals to $p(x)$. Hence, we have

$$-\sum_{x,y} p(x,y)\log(p(x)) = -\sum_{x} p(x)\log(p(x)).$$

In a similar manner, we can also write

$$-\sum_{x,y} p(x,y)\log(p(y)) = -\sum_{y} p(y)\log(p(y)).$$

Example 1.20 Prove the following property

$$H(\tilde{X},\tilde{Y}) \leq H(\tilde{X}) + H(\tilde{Y}).$$

Solution 1.20 Using the definition of entropy in

$$H(\tilde{X}) + H(\tilde{Y})$$

we get

$$H(\tilde{X}) + H(\tilde{Y}) = -\sum_{x} p(x)\log(p(x)) - \sum_{y} p(y)\log(p(y))$$

where the summation expressions can be written as

$$H(\tilde{X}) + H(\tilde{Y}) = -\sum_{x,y} p(x,y)\log(p(x)) - \sum_{x,y} p(x,y)\log(p(y)). \tag{1.57}$$

Combining the logarithmic term on the right hand side of (1.57), we get

$$H(\tilde{X})+H(\tilde{Y}) = -\sum_{x,y} p(x,y)\log(p(x)p(y)). \tag{1.58}$$

Employing the log-sum inequality

$$\sum_i p_i \log p_i \geq \sum_i p_i \log q_i$$

which can be written also as

$$-\sum_i p_i \log q_i \geq -\sum_i p_i \log p_i$$

for the right side of the (1.58), we obtain

$$H(\tilde{X})+H(\tilde{Y}) = -\sum_{x,y} p(x,y)\log(p(x)p(y)) \geq \underbrace{-\sum_{x,y} p(x,y)\log(p(x,y))}_{H(\tilde{X},\tilde{Y})}.$$

Hence, we have

$$H(\tilde{X},\tilde{Y}) \leq H(\tilde{X})+H(\tilde{Y}).$$

Example 1.21 Prove the following property

$$H(\tilde{X}|\tilde{Y}) \leq H(\tilde{X}).$$

Solution 1.21 In our previous example, we have shown that

$$H(\tilde{X},\tilde{Y}) \leq H(\tilde{X})+H(\tilde{Y})$$

which can be written as

$$H(\tilde{X},\tilde{Y}) - H(\tilde{Y}) \leq H(\tilde{X}) \tag{1.59}$$

where substituting the definitions for the joint and marginal entropies, we obtain

$$-\sum_{x,y} p(x,y)\log(p(x,y)) + \sum_y p(y)\log(p(y)) \leq -\sum_x p(x)\log(p(x))$$

in which replacing

$$\sum_{y} p(y)\log(p(y))$$

by

$$\sum_{x,y} p(x,y)\log(p(y))$$

we get

$$-\sum_{x,y} p(x,y)\log(p(x,y)) + \sum_{x,y} p(x,y)\log(p(y)) \leq -\sum_{x} p(x)\log(p(x)). \quad (1.60)$$

Grouping the common terms $p(x,y)$ in (1.60), we obtain

$$-\sum_{x,y} p(x,y)\log\left(\frac{p(x,y)}{p(y)}\right) \leq -\sum_{x} p(x)\log(p(x))$$

which can also be written as

$$\underbrace{-\sum_{x,y} p(x,y)\log(p(x|y))}_{H(\tilde{X}|\tilde{Y})} \leq \underbrace{-\sum_{x} p(x)\log(p(x))}_{H(\tilde{X})}$$

where the left hand side corresponds to $H(\tilde{X}|\tilde{Y})$, and the right hand side corresponds to $H(\tilde{X})$. Thus, we get

$$H(\tilde{X}|\tilde{Y}) \leq H(\tilde{X}).$$

We started with (1.59) and showed that the left hand side of (1.59) equals to $H(\tilde{X}|\tilde{Y})$, thus we can also write this result as

$$H(\tilde{X}|\tilde{Y}) = H(\tilde{X},\tilde{Y}) - H(\tilde{Y}).$$

Remark $H(\tilde{X}|\tilde{Y})$ can be interpreted as the amount of remaining uncertainty about the random variable $\tilde{X}$ when all the elements of the random variable $\tilde{Y}$ are known, or it can be interpreted as the amount of required information to identify the random variable $\tilde{X}$ when all the elements of the random variable $\tilde{Y}$ are known, or it can be interpreted as the amount of different information supplied by the random variable $\tilde{X}$ when the random variable $\tilde{Y}$ is known.

Let's summarize the formulas we obtained up to now.

Summary

$$H(\tilde{X},\tilde{Y}) \leq H(\tilde{X}) + H(\tilde{Y})$$
$$H(\tilde{X}|\tilde{Y}) \leq H(\tilde{X})$$
$$H(\tilde{X}|\tilde{Y}) = H(\tilde{X},\tilde{Y}) - H(\tilde{Y})$$

The above formulas can also be described using Venn diagram as shown in Fig. 1.9.

Example 1.22 For the discrete memoryless channel shown in Fig. 1.10, the source symbol probabilities are given as

$$p(x_1) = \alpha \qquad p(x_2) = 1 - \alpha.$$

Find the conditional entropy $H(\tilde{X}|\tilde{Y})$.

Solution 1.22 The conditional entropy $H(\tilde{X}|\tilde{Y})$ can be calculated using either

$$H(\tilde{X}|\tilde{Y}) = -\sum_{x,y} p(x,y)\log(p(x|y)) \tag{1.61}$$

or using the formula pairs

$$\begin{aligned} H(\tilde{X}|\tilde{Y}) &= \sum_{y} p(y)H(\tilde{X}|y) \\ H(\tilde{X}|y) &= -\sum_{x} p(x|y)\log(p(x|y)). \end{aligned} \tag{1.62}$$

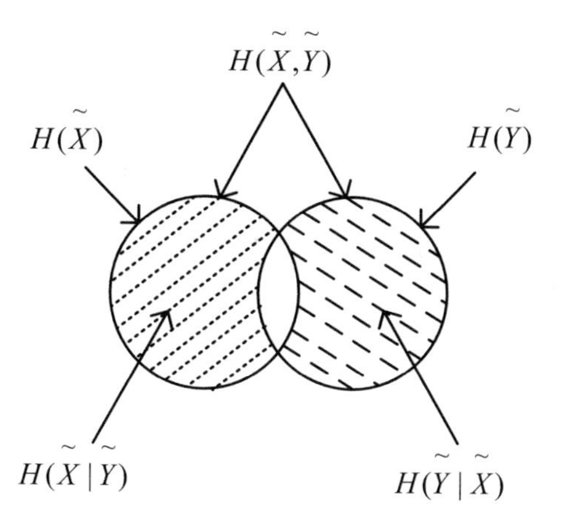

Fig. 1.9 Venn diagram illustration of the entropy

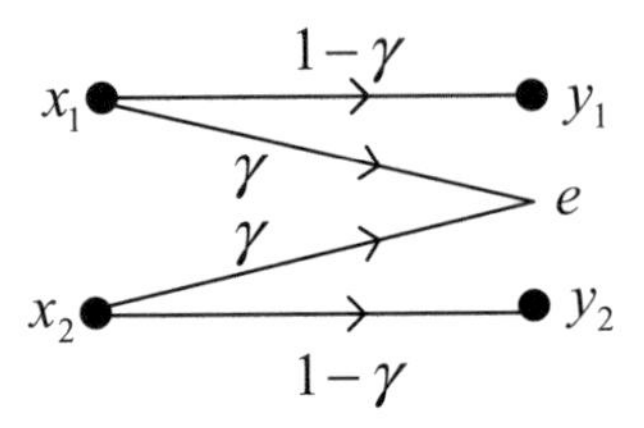

Fig. 1.10 Discrete communication channel for Example 1.22

If we use (1.61) directly, we need to calculate $p(x|y)$ for every x, y pairs. However, in first equation of (1.62), $H(\tilde{X}|y)$ expression appears, and this expression can be evaluated logically for some output symbols without doing any calculation. For this reason, it is logical to employ (1.62) for the conditional entropy calculation. Let's first show the binary erasure channel for each symbol output in detail as shown in Fig. 1.11.

If we consider the conditional entropies $H(\tilde{X}|y)$ for each channel output as shown in Fig. 1.11, we can conclude that

$$H(\tilde{X}|y_1) = 0 \qquad H(\tilde{X}|y_2) = 0.$$

Since if y_1 or y_2 are known, there is no uncertainty left about $\tilde{X}$, i.e., if y_1 is received, then we can estimate the transmitted symbol, x_1 in this case, and similarly if y_2 is received, then we can estimate the transmitted symbol x_2 without any doubt, and when we expand the formula in (1.62), we get

$$\begin{aligned} &H(\tilde{X}|\tilde{Y}) = \sum_y p(y)H(\tilde{X}|y) \rightarrow \\ &H(\tilde{X}|\tilde{Y}) = p(y_1)\underbrace{H(\tilde{X}|y_1)}_{=0} + p(e)H(\tilde{X}|e) + p(y_2)\underbrace{H(\tilde{X}|y_2)}_{=0} \rightarrow \\ &H(\tilde{X}|\tilde{Y}) = p(e)H(\tilde{X}|e) \end{aligned} \tag{1.63}$$

where it is seen that the calculation of $H(\tilde{X}|\tilde{Y})$ can be achieved by finding $p(e)$ and $H(\tilde{X}|e)$. The probability $p(e)$ can be calculated as

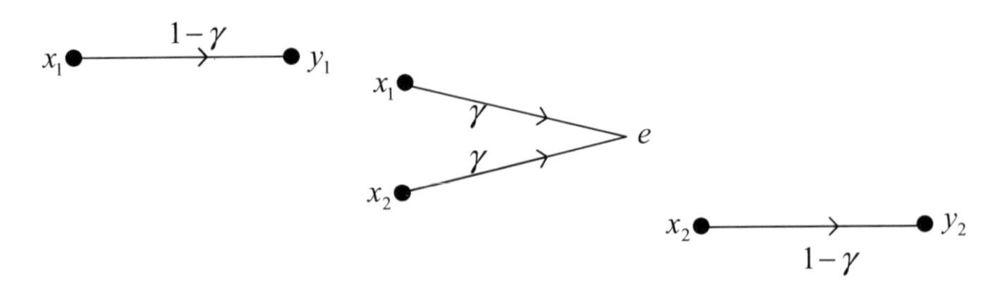

Fig. 1.11 Binary erasure channel in details

$$p(e) = \sum_x p(e,x) \rightarrow p(e) = \sum_x p(e|x)p(x)$$

leading to

$$p(e) = p(e|x_1)p(x_1) + p(e|x_2)p(x_2)$$

which is evaluated as

$$p(e) = \gamma\alpha + \gamma(1-\alpha) \rightarrow p(e) = \gamma. \tag{1.64}$$

The conditional entropy $H(\tilde{X}|e)$ can be calculated using

$$H(\tilde{X}|e) = -\sum_x p(x|e)\log(p(x|e)). \tag{1.65}$$

Expanding (1.65), we obtain

$$\begin{aligned} H(\tilde{X}|e) &= -\sum_x p(x|e)\log(p(x|e)) \rightarrow \\ H(\tilde{X}|e) &= -[p(x_1|e)\log(p(x_1|e)) + p(x_2|e)\log(p(x_2|e))]. \end{aligned} \tag{1.66}$$

Since,

$$p(x|e) = \frac{p(x,e)}{p(e)} \rightarrow p(x|e) = \frac{p(e|x)p(x)}{p(e)}$$

Equation (1.66) can be written as

$$H(\tilde{X}|e) = -\left[\frac{p(e|x_1)}{p(e)}p(x_1)\log\left(\frac{p(e|x_1)}{p(e)}p(x_1)\right) + \frac{p(e|x_2)}{p(e)}p(x_2)\log\left(\frac{p(e|x_2)}{p(e)}p(x_2)\right)\right]$$

where inserting the given values in the question, (1.66) can be evaluated as

$$H(\tilde{X}|e) = -\left[\frac{\gamma}{\gamma}\alpha\log\left(\frac{\gamma}{\gamma}\alpha\right) + \frac{\gamma}{\gamma}(1-\alpha)\log\left(\frac{\gamma}{\gamma}(1-\alpha)\right)\right]$$

which is simplified as

$$H(\tilde{X}|e) = -[\alpha\log(\alpha) + (1-\alpha)\log((1-\alpha))]. \tag{1.67}$$

The right hand side of (1.67) is a function of α only, and this special expression can be denoted by $H_b(\alpha)$. Thus

$$H(\tilde{X}|e) = H_b(\alpha). \tag{1.68}$$

And finally, combining the results in (1.63), (1.64) and (1.68), we obtain

$$H(\tilde{X}|\tilde{Y}) = \gamma H_b(\alpha). \tag{1.69}$$

The graph of the concave function $H_b(\alpha)$ is shown in Fig. 1.1 where it is seen that the function gets its maximum value when $\alpha = 0.5$ and it gets it minimum value 0 when $\alpha = 0$ or $\alpha = 1$. When (1.69) is inspected, it is seen that as α approaches to 0 or 1, the value of $H_b(\alpha)$ decreases and this results also in a decrement in the conditional entropy $H(\tilde{X}|\tilde{Y})$. This means that when $\tilde{Y}$ is known, if the randomness of $\tilde{X}$ decreases, i.e., the probability of sending a symbol becomes significant considering the other, then $H(\tilde{X}|\tilde{Y})$ decreases as well.

Example 1.23 For the binary erasure channel given in the previous example, find $H(\tilde{Y}|\tilde{X})$.

Solution 1.23 The conditional entropy $H(\tilde{Y}|\tilde{X})$ can be calculated using either

$$H(\tilde{Y}|\tilde{X}) = -\sum_{x,y} p(x,y)\log(p(y|x)) \tag{1.70}$$

or using the formula pair

$$\begin{aligned} H(\tilde{Y}|\tilde{X}) &= \sum_{x} p(x) H(\tilde{Y}|x) \\ H(\tilde{Y}|x) &= -\sum_{y} p(y|x)\log(p(y|x)). \end{aligned} \tag{1.71}$$

Let's use the formula pair in (1.71). The binary erasure channel for a given transmitted symbol is shown in details in Fig. 1.12. $H(\tilde{Y}|x_1)$ can be calculated as

$$H(\tilde{Y}|x_1) = -\sum_{y} p(y|x_1)\log(p(y|x_1)) \tag{1.72}$$

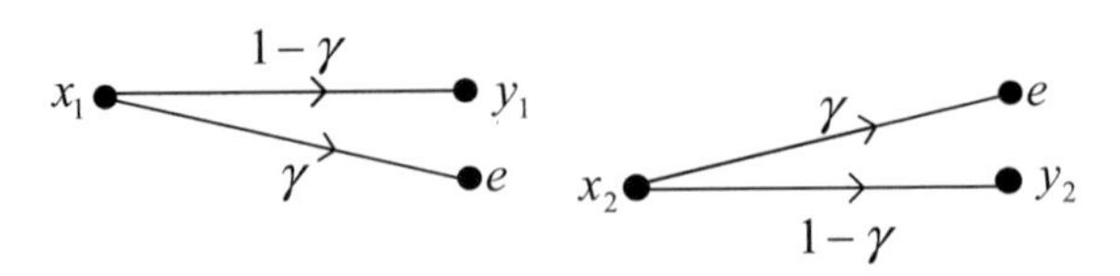

Fig. 1.12 Binary erasure channel in two parts

When (1.72) is expanded, we obtain

$$H(\tilde{Y}|x_1) = -\left[\underbrace{p(y_1|x_1)}_{1-\gamma}\log\left(\underbrace{p(y_1|x_1)}_{1-\gamma}\right) + \underbrace{p(e|x_1)}_{\gamma}\log\left(\underbrace{p(e|x_1)}_{\gamma}\right)\right]$$

where the right hand side equals to $H_b(\gamma)$, hence, $H(\tilde{Y}|x_1) = H_b(\gamma)$. In a similar manner, $H(\tilde{Y}|x_2)$ is calculated as

$$H(\tilde{Y}|x_2) = -\left[\underbrace{p(y_2|x_2)}_{1-\gamma}\log\left(\underbrace{p(y_2|x_2)}_{1-\gamma}\right) + \underbrace{p(e|x_2)}_{\gamma}\log\left(\underbrace{p(e|x_2)}_{\gamma}\right)\right]$$

where the right hand side equals to $H_b(\gamma)$, hence, $H(\tilde{Y}|x_2) = H_b(\gamma)$. Using the first equation of (1.71), the conditional entropy can be calculated as

$$H(\tilde{Y}|\tilde{X}) = \sum_x p(x)H(\tilde{Y}|x) \rightarrow H(\tilde{Y}|\tilde{X}) = \underbrace{p(x_1)}_{\alpha}\underbrace{H(\tilde{Y}|x_1)}_{H_b(\gamma)} + \underbrace{p(x_2)}_{1-\alpha}\underbrace{H(\tilde{Y}|x_2)}_{H_b(\gamma)}$$

which can be simplified as

$$H(\tilde{Y}|\tilde{X}) = \underbrace{p(x_1)}_{\alpha}\underbrace{H(\tilde{Y}|x_1)}_{H_b(\gamma)} + \underbrace{p(x_2)}_{1-\alpha}\underbrace{H(\tilde{Y}|x_2)}_{H_b(\gamma)} \rightarrow H(\tilde{Y}|\tilde{X}) = H_b(\gamma).$$

Exercise For the discrete memoryless channel shown in Fig. 1.13, the source symbol probabilities are given as

$$p(x_1) = \alpha \qquad p(x_2) = 1 - \alpha.$$

Find the conditional entropy $H(\tilde{X}|\tilde{Y})$.

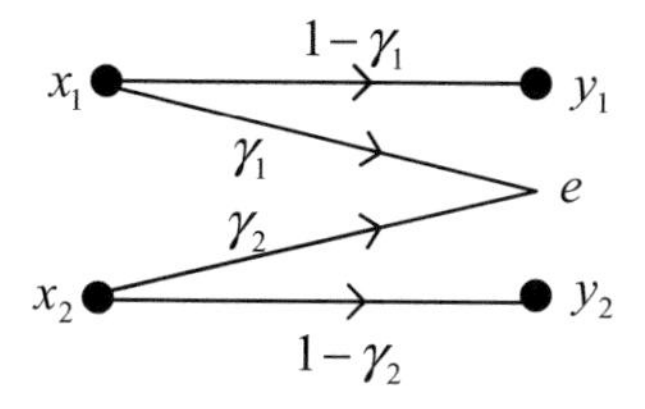

Fig. 1.13 Binary erasure channel for exercise

1.4 Information Channels

When we ask the question, what a channel is, to an ordinary people outside, he or she consider the channel as a water channel or a tunnel. However, in information theory a communication channel is nothing but a set of probabilities among transmitted and received symbols. The probability of receiving an output symbol for any input symbol describes the information channel. Since there is more than one input symbol to be transmitted and more than one candidate output symbol at the receiver side, the information channel is usually described by a probability matrix. The source and destination parts of a communication system can be described by random variables as depicted in Fig. 1.14.

A discrete memoryless information channel can also be graphically illustrated. In this case, channel transition probabilities are shown along the lines. As an example, in Fig. 1.15, channel transition probabilities are indicated along the lines.

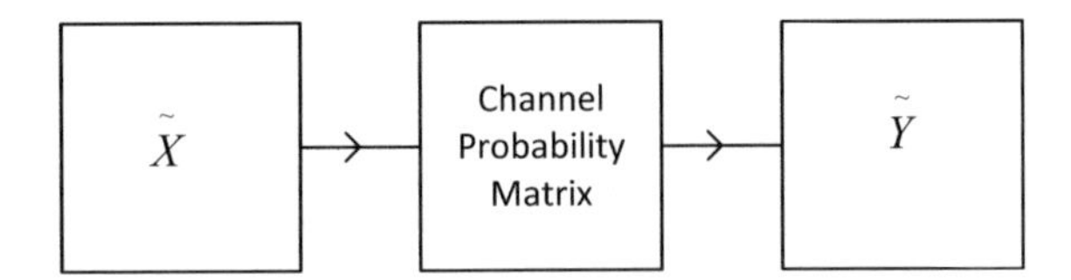

Fig. 1.14 Communication system

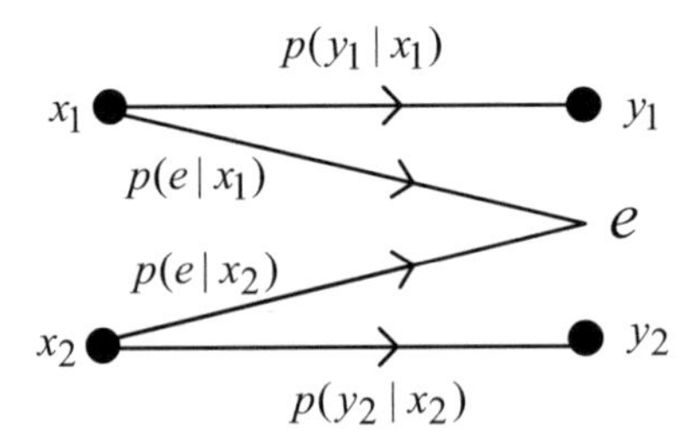

Fig. 1.15 Binary erasure channel with transition probabilities

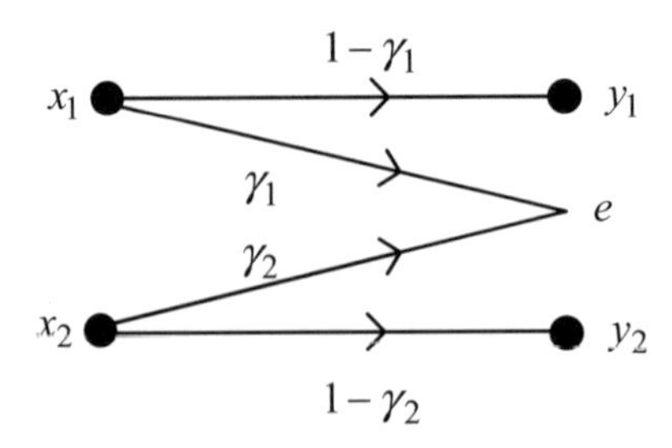

Fig. 1.16 Binary erasure channel for Example 1.24

Example 1.24 The binary erasure channel given in Fig. 1.16 can be described by the matrix

$$P = \begin{array}{c|c} x_1 & \left[\begin{array}{ccc} 1-\gamma_1 & \gamma_1 & 0 \\ 0 & \gamma_2 & 1-\gamma_2 \end{array}\right] \\ x_2 & \\ \hline & y_1 \quad y_2 \quad y_3 \end{array}$$

Example 1.25 The discrete random variables $\tilde{X}$, $\tilde{Y}$ and $\tilde{Z}$ have the range sets $R_{\tilde{X}} = \{x_1, x_2\}$, $R_{\tilde{Y}} = \{y_1, y_2\}$, and $R_{\tilde{Z}} = \{z_1, z_2\}$. Find $H(\tilde{X}|\tilde{Y}, z)$.

Solution 1.25 The conditional entropy $H(\tilde{X}|\tilde{Y}, z)$ can be calculated using

$$H(\tilde{X}|\tilde{Y}, z) = -\sum_{x,y} p(x, y, z) \log(p(x|y, z))$$

which can be calculated for z_1 and z_2 separately as

$$H(\tilde{X}|\tilde{Y}, z_1) = -\sum_{x,y} p(x, y, z_1) \log(p(x|y, z_1)) \tag{1.73}$$

and

$$H(\tilde{X}|\tilde{Y}, z_2) = -\sum_{x,y} p(x, y, z_2) \log(p(x|y, z_2)). \tag{1.74}$$

When (1.73) is expanded for all x and y pairs, we obtain

$$\begin{aligned} H(\tilde{X}|\tilde{Y}, z_1) = -[& p(x_1, y_1, z_1) \log(p(x_1|y_1, z_1)) + p(x_1, y_2, z_1) \log(p(x_1|y_2, z_1)) \\ & + p(x_2, y_1, z_1) \log(p(x_2|y_1, z_1)) + p(x_2, y_2, z_1) \log(p(x_2|y_2, z_1))]. \end{aligned}$$

Similarly, when (1.74) is expanded for all x and y pairs, we obtain

$$\begin{aligned} H(\tilde{X}|\tilde{Y}, z_2) = -[& p(x_1, y_1, z_2) \log(p(x_1|y_1, z_2)) + p(x_1, y_2, z_2) \log(p(x_1|y_2, z_2)) \\ & + p(x_2, y_1, z_2) \log(p(x_2|y_1, z_2)) + p(x_2, y_2, z_2) \log(p(x_2|y_2, z_2))]. \end{aligned}$$

Example 1.26 The discrete random variables $\tilde{X}$, $\tilde{Y}$ and $\tilde{Z}$ have the range sets $R_{\tilde{X}} = \{x_1, x_2\}$, $R_{\tilde{Y}} = \{y_1, y_2\}$ and $R_{\tilde{Z}} = \{z_1, z_2\}$. Write an expression for $H(\tilde{X}|\tilde{Y}, \tilde{Z})$ in terms of the conditional entropies $H(\tilde{X}|\tilde{Y}, z_1)$ and $H(\tilde{X}|\tilde{Y}, z_2)$.

Solution 1.26 The conditional entropy $H(\tilde{X}|\tilde{Y},\tilde{Z})$ can be calculated using

$$H(\tilde{X}|\tilde{Y},\tilde{Z}) = \sum_{z} p(z)H(\tilde{X}|\tilde{Y},z). \tag{1.75}$$

When (1.75) is expanded for the range set of $\tilde{Z}$, we obtain

$$H(\tilde{X}|\tilde{Y},\tilde{Z}) = p(z_1)H(\tilde{X}|\tilde{Y},z_1) + p(z_2)H(\tilde{X}|\tilde{Y},z_2).$$

Note: The conditional expression $H(\tilde{X}|\tilde{Y},\tilde{Z})$ can also be evaluated using

$$H(\tilde{X}|\tilde{Y},\tilde{Z}) = \sum_{y} p(y)H(\tilde{X}|\tilde{Z},y)$$

or using

$$H(\tilde{X}|\tilde{Y},\tilde{Z}) = \sum_{y,z} p(y,z)H(\tilde{X}|y,z).$$

Example 1.27 The discrete random variables $\tilde{X}$, $\tilde{Y}$ and $\tilde{Z}$ have the range sets $R_{\tilde{X}} = \{x_1, x_2\}$ and $R_{\tilde{Y}} = \{y_1, y_2\}$, $R_{\tilde{Z}} = \{z_1, z_2\}$. Write an expression for the calculation of $H(\tilde{X}|y_1, z_2)$.

Solution 1.27 The conditional entropy $H(\tilde{X}|y_1, z_2)$ can be calculated using

$$H(\tilde{X}|y_1, z_2) = -\sum_{x} p(x|y_1, z_2) \log(p(x|y_1, z_2)) \tag{1.76}$$

When (1.76) is expanded, we obtain

$$H(\tilde{X}|y_1, z_2) = -[p(x_1|y_1, z_2) \log(p(x_1|y_1, z_2)) + p(x_2|y_1, z_2) \log(p(x_2|y_1, z_2))] \tag{1.77}$$

Example 1.28 $\tilde{X}$ is a discrete random variable and the random variable $\tilde{Y}$ is defined as $\tilde{Y} = \tilde{X}^2$. Calculate $H(\tilde{Y})$.

Solution 1.28 We can calculate $H(\tilde{Y})$ using

$$H(\tilde{Y}) = \sum_{y} p(y) \log(p(y))$$

where $p(y)$ can be written as

$$p(y) = Prob(\tilde{Y} = y) \rightarrow$$
$$p(y) = Prob(\tilde{X}^2 = y) \rightarrow$$
$$p(y) = Prob(\tilde{X} = \sqrt{y}) + Prob(\tilde{X} = -\sqrt{y})$$
$$p_{\tilde{Y}}(y) = p_{\tilde{X}}(\sqrt{y}) + p_{\tilde{X}}(-\sqrt{y})$$

then, $H(\tilde{Y})$ can be written as

$$H(\tilde{Y}) = \sum_y (p_{\tilde{X}}(\sqrt{y}) + p_{\tilde{X}}(-\sqrt{y})) \log(p_{\tilde{X}}(\sqrt{y}) + p_{\tilde{X}}(-\sqrt{y}))$$

Example 1.29 For the discrete random variable $\tilde{X}$, the range set is given as $R_{\tilde{X}} = \{x_1, x_2, x_3\}$, and for the discrete random variable $\tilde{Y}$, the range set is given as $R_{\tilde{Y}} = \{x_1, y_2, y_3\}$. Calculate the conditional entropy

$$H(\tilde{X}|\tilde{Y} = x_1).$$

Solution 1.29 In this example, the range sets $R_{\tilde{X}}$ and $R_{\tilde{Y}}$ have a common element x_1. The conditional entropy in its general expression $H(\tilde{X}|\tilde{Y} = y)$ is calculated as

$$H(\tilde{X}|\tilde{Y} = y) = -\sum_x p(x|y) \log(x|y).$$

For the conditional entropy $H(\tilde{X}|\tilde{Y} = x_1)$ given in the question, the calculation can be performed in a similar manner as

$$H(\tilde{X}|\tilde{Y} = x_1) = -\sum_x p(x|x_1) \log(x|x_1). \tag{1.78}$$

When (1.78) is expanded, we obtain

$$\begin{aligned} H(\tilde{X}|\tilde{Y} = x_1) = -[&p(x_1|x_1) \log(x_1|x_1) + p(x_2|x_1) \log(x_2|x_1) \\ &+ p(x_3|x_1) \log(x_3|x_1)] \end{aligned} \tag{1.79}$$

where $p(x_1|x_1)$ can be calculated as

$$p(x_1|x_1) = \frac{p(x_1, x_1)}{p(x_1)} \rightarrow p(x_1|x_1) = \frac{p(x_1)}{p(x_1)} \rightarrow p(x_1|x_1) = 1$$

Then, (1.79) reduces to

$$H(\tilde{X}|\tilde{Y} = x_1) = -[p(x_2|x_1) \log(x_2|x_1) + p(x_3|x_1) \log(x_3|x_1)] \tag{1.80}$$

For any random variable, the elements of the range set corresponds to some disjoint sets, for this reason

$$p(x_2|x_1) = p(x_2) \qquad p(x_3|x_1) = p(x_3).$$

Then, (1.80) can be written as

$$H(\tilde{X}|\tilde{Y} = x_1) = -[p(x_2)\log(p(x_2)) + p(x_3)\log(p(x_3))]. \tag{1.81}$$

If we calculate $H(\tilde{X})$, we find it as

$$H(\tilde{X}) = -[p(x_1)\log(p(x_1)) + p(x_2)\log(p(x_2)) + p(x_3)\log(p(x_3))]. \tag{1.82}$$

When (1.81) and (1.82) are inspected, we see that if an elements of the range set of the discrete random variable is known by another random variable, the conditional entropy gets a lower value. This is clearly seen from the right hand side of (1.81).

Exercise For the discrete random variable $\tilde{X}$, the range set is given as $R_{\tilde{X}} = \{x_1, x_2, x_3\}$. For the discrete random variable $\tilde{Y}$, the range set is given as $R_{\tilde{Y}} = \{x_1, x_2, y_3\}$. Calculate the conditional entropy

$$H(\tilde{X}|\tilde{Y}).$$

1.5 Mutual Information

The mutual information between two random variables $\tilde{X}$ and $\tilde{Y}$ is defined as

$$I(\tilde{X};\tilde{Y}) = H(\tilde{X}) - H(\tilde{X}|\tilde{Y}) \text{ bits/symbol.} \tag{1.83}$$

It is clear from (1.83) that the unit of the mutual information is bits/symbol.
Note:

$$\sum_{x,y} p(x,y)\log(p(x)) = \sum_{x} p(x)\log(p(x)).$$

If the mathematical expressions of $H(\tilde{X})$ and $H(\tilde{X}|\tilde{Y})$ are substituted into (1.83), we obtain

$$I(\tilde{X};\tilde{Y}) = -\sum_{x} p(x)\log(p(x)) + \sum_{x,y} p(x,y)\log(p(x|y)) \tag{1.84}$$

in which substituting

$$\sum_{x,y} p(x,y)\log(p(x))$$

for

$$\sum_{x} p(x)\log(p(x))$$

we get

$$I(\tilde{X};\tilde{Y}) = -\sum_{x,y} p(x,y)\log(p(x)) + \sum_{x,y} p(x,y)\log(p(x|y)). \tag{1.85}$$

In (1.85), $p(x,y)$ is common term for summations, and taking the common term out and combining the logarithmic terms, we get

$$I(\tilde{X};\tilde{Y}) = \sum_{x,y} p(x,y)\log\left(\frac{p(x|y)}{p(x)}\right)$$

where replacing

$$p(x|y)$$

by

$$\frac{p(x,y)}{p(y)}$$

we obtain

$$I(\tilde{X};\tilde{Y}) = \sum_{x,y} p(x,y)\log\left(\frac{p(x,y)}{p(x)p(y)}\right). \tag{1.86}$$

The joint probability mass function $p(x,y)$ in (1.86) can also we written as $p(x,y) = p(y|x)p(x)$, in this case the mutual information expression in (1.86) takes the form

$$I(\tilde{X};\tilde{Y}) = \sum_{x,y} p(x,y)\log\left(\frac{p(y|x)}{p(y)}\right). \tag{1.87}$$

To sum it up, the mutual information for two discrete random variables $\tilde{X}$ and $\tilde{Y}$ can be expressed in one of these there forms

$$I(\tilde{X};\tilde{Y}) = \sum_{x,y} p(x,y)\log\left(\frac{p(x|y)}{p(x)}\right) \quad I(\tilde{X};\tilde{Y}) = \sum_{x,y} p(x,y)\log\left(\frac{p(y|x)}{p(y)}\right)$$
$$I(\tilde{X};\tilde{Y}) = \sum_{x,y} p(x,y)\log\left(\frac{p(x,y)}{p(x)p(y)}\right). \tag{1.88}$$

At the beginning of this section, we defined mutual information in terms of marginal and conditional entropies. An equivalent definition to (1.83) can be given as

$$I(\tilde{X};\tilde{Y}) = H(\tilde{Y}) - H(\tilde{Y}|\tilde{X}) \text{ bits/symbol.} \tag{1.89}$$

If we put the mathematical expressions of marginal and conditional entropies in (1.89), we obtain the same formula (1.88). Using the identities

$$H(\tilde{X}|\tilde{Y}) = H(\tilde{X},\tilde{Y}) - H(\tilde{Y})$$

or

$$H(\tilde{Y}|\tilde{X}) = H(\tilde{X},\tilde{Y}) - H(\tilde{X})$$

in (1.83) or in (1.89) respectively, we obtain the alternative form of the mutual information as

$$I(\tilde{X};\tilde{Y}) = H(\tilde{X}) + H(\tilde{Y}) - H(\tilde{X},\tilde{Y}) \text{ bits/symbol.}$$

To sum it up, the mutual information for two discrete random variables $\tilde{X}$ and $\tilde{Y}$ can be expressed in terms of the marginal, conditional and joint mutual information as in one of these there forms

$$I(\tilde{X};\tilde{Y}) = H(\tilde{X}) - H(\tilde{X}|\tilde{Y}) \qquad I(\tilde{X};\tilde{Y}) = H(\tilde{Y}) - H(\tilde{Y}|\tilde{X})$$
$$I(\tilde{X};\tilde{Y}) = H(\tilde{X}) + H(\tilde{Y}) - H(\tilde{X},\tilde{Y}).$$

The relationship among $I(\tilde{X};\tilde{Y}), H(\tilde{X}), H(\tilde{Y}), H(\tilde{X}|\tilde{Y}), H(\tilde{Y}|\tilde{X})$ and $H(\tilde{X},\tilde{Y})$ can be described by Venn diagram as shown in Fig. 1.17.

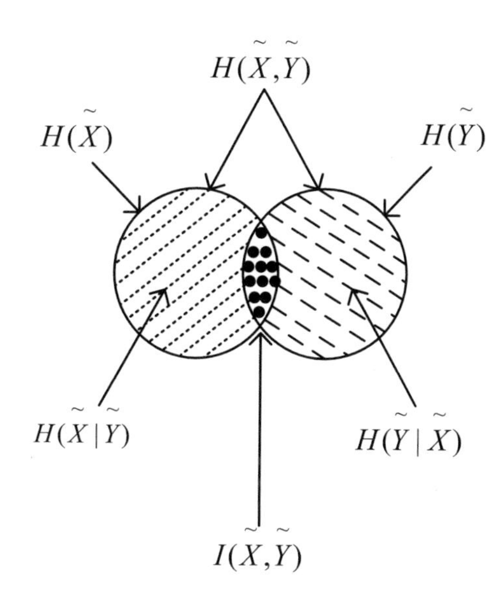

Fig. 1.17 Venn diagram illustration of the mutual information

Example 1.30 Find the mutual information between input and output of the binary erasure channel shown in Fig. 1.18, assume that $p(x_1) = \alpha$ and $p(x_2) = 1 - \alpha$.

Solution 1.30 The mutual information $I(\tilde{X};\tilde{Y})$ can be calculated using

$$I(\tilde{X};\tilde{Y}) = H(\tilde{X}) - H(\tilde{X}|\tilde{Y}) \tag{1.90}$$

The source entropy $H(\tilde{X})$ can be found as

$$H(\tilde{X}) = -[\alpha \log a + (1-\alpha)\log(1-\alpha)] \rightarrow H(\tilde{X}) = H_b(\alpha). \tag{1.91}$$

In Example 1.22, we calculated conditional entropy $H(\tilde{X}|\tilde{Y})$ of the binary erasure channel as

$$H(\tilde{X}|\tilde{Y}) = \gamma H_b(\alpha). \tag{1.92}$$

Substituting (1.91) and (1.92) into (1.90), we get

$$I(\tilde{X};\tilde{Y}) = (1-\gamma)H_b(\alpha). \tag{1.93}$$

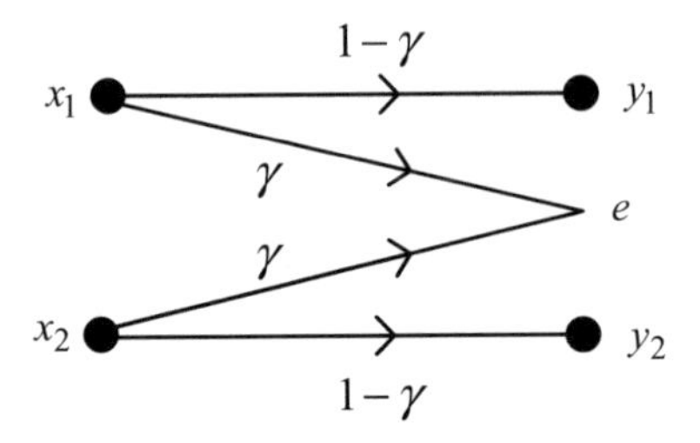

Fig. 1.18 Binary erasure channel for Example 1.30

Example 1.31 For the discrete communication channel shown in Fig. 1.19, the input symbol probabilities and transition probabilities are given as

$$p(x_1) = \frac{1}{8} \qquad p(x_2) = \frac{7}{8}$$

$$p(y_1|x_1) = \frac{1}{4} \qquad p(y_1|x_2) = \frac{1}{8} \qquad p(y_2|x_1) = \frac{3}{4} \qquad p(y_2|x_2) = \frac{7}{8}.$$

Find $I(\tilde{X};\tilde{Y})$.

Solution 1.31 The mutual information $I(\tilde{X};\tilde{Y})$ can be calculated using

$$I(\tilde{X};\tilde{Y}) = \sum_{x,y} p(x,y) \log\left(\frac{p(x,y)}{p(x)p(y)}\right)$$

where the joint and marginal probabilities $p(x,y)$, $p(y)$ can be calculated using

$$p(x,y) = p(y|x)p(x)p(y) = \sum_{x} p(x,y)$$

as in

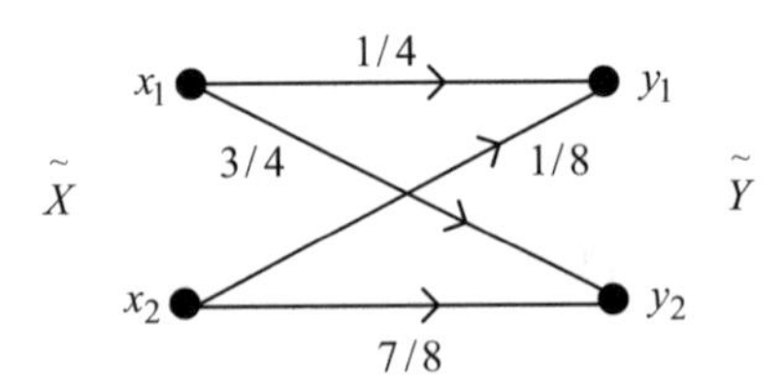

Fig. 1.19 Discrete communication channel for Example 1.31

$$p(x = x_1, y = y_1) = p(y = y_1|x = x_1)p(x = x_1) \rightarrow p(x = x_1, y = y_1) = \frac{1}{32}$$

$$p(x = x_1, y = y_2) = p(y = y_2|x = x_1)p(x = x_1) \rightarrow p(x = x_1, y = y_2) = \frac{3}{32}$$

$$p(x = x_2, y = y_1) = p(y = y_1|x = x_2)p(x = x_2) \rightarrow p(x = x_2, y = y_1) = \frac{7}{64}$$

$$p(x = x_2, y = y_2) = p(y = y_2|x = x_2)p(x = x_2) \rightarrow p(x = x_2, y = y_2) = \frac{49}{64}$$

$$p(y = y_1) = \frac{9}{64} \qquad p(y = y_2) = \frac{55}{64}.$$

Then, we can calculate the mutual information as

$$I(\tilde{X}; \tilde{Y}) = p(x_1, y_1) \log\left(\frac{p(x_1, y_1)}{p(x_1)p(y_1)}\right) + p(x_1, y_2) \log\left(\frac{p(x_1, y_2)}{p(x_1)p(y_2)}\right)$$
$$+ p(x_2, y_1) \log\left(\frac{p(x_2, y_1)}{p(x_2)p(y_1)}\right) + p(x_2, y_2) \log\left(\frac{p(x_2, y_2)}{p(x_2)p(y_2)}\right)$$

leading to

$$I(\tilde{X}; \tilde{Y}) = \frac{1}{32} \log\left(\frac{\frac{1}{32}}{\frac{1}{8} \times \frac{9}{64}}\right) + \frac{3}{32} \log\left(\frac{\frac{3}{32}}{\frac{1}{8} \times \frac{55}{64}}\right)$$
$$+ \frac{7}{64} \log\left(\frac{\frac{7}{64}}{\frac{7}{8} \times \frac{9}{64}}\right) + \frac{49}{64} \log\left(\frac{\frac{49}{64}}{\frac{7}{8} \times \frac{55}{64}}\right)$$

which is evaluated as

$$I(\tilde{X}; \tilde{Y}) = 0.0088.$$

Exercise For the discrete communication channel shown in Fig. 1.20, calculate $I(\tilde{X}; \tilde{Y})$.

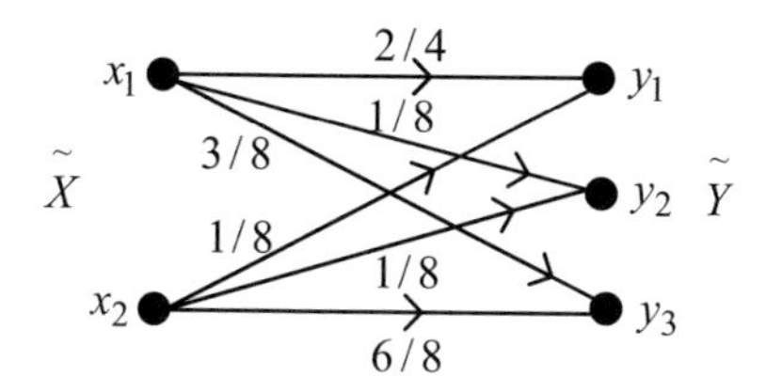

Fig. 1.20 Discrete communication channel for exercise

1.5.1 Properties of the Mutual Information

(a) Symmetry property states that

$$I(\tilde{X};\tilde{Y}) = I(\tilde{Y};\tilde{X}).$$

(b) Mutual information is a nonnegative quantity, i.e.,

$$I(\tilde{X};\tilde{Y}) \geq 0.$$

The proof of this property is straight forward. Since

$$H(\tilde{X}|\tilde{Y}) \leq H(\tilde{X})$$

and

$$I(\tilde{X};\tilde{Y}) = H(\tilde{X}) - H(\tilde{X}|\tilde{Y})$$

it is obvious that

$$I(\tilde{X};\tilde{Y}) \geq 0.$$

Summary Various expressions for mutual information in terms of entropy and conditional entropy can be written. Let's summarize those expressions as follows. We know that mutual information can be expressed using either

$$I(\tilde{X};\tilde{Y}) = H(\tilde{X}) - H(\tilde{X}|\tilde{Y}) \tag{1.94}$$

or

$$I(\tilde{X};\tilde{Y}) = H(\tilde{Y}) - H(\tilde{Y}|\tilde{X}) \tag{1.95}$$

Substituting

$$H(\tilde{X}|\tilde{Y}) = H(\tilde{X},\tilde{Y}) - H(\tilde{Y})$$

or

$$H(\tilde{Y}|\tilde{X}) = H(\tilde{X},\tilde{Y}) - H(\tilde{X})$$

into (1.94) or (1.95) respectively, the mutual information is expressed as

$$I(\tilde{X};\tilde{Y}) = H(\tilde{X}) + H(\tilde{Y}) - H(\tilde{X},\tilde{Y}).$$

Example 1.32 Show that

$$H(\tilde{X}) \leq \log|R_{\tilde{X}}|$$

where $|R_{\tilde{X}}|$ is the number of elements in the range set of the discrete random variable $\tilde{X}$.

Solution 1.32 Before proceeding with the solution, let's refresh our calculus knowledge for logarithm function. We have

$$\log_a b = \frac{\log_e b}{\log_e a} \rightarrow \log_a b = \frac{\ln b}{\ln a} \tag{1.96}$$

Using (1.96), we can write

$$\log_2 x = \frac{\ln x}{\ln 2}. \tag{1.97}$$

In addition, for the $\ln(\cdot)$ function, we have the property

$$1 - \frac{1}{x} \leq \ln x \leq x - 1 \tag{1.98}$$

where $x > 0$.

Now, we can start to our proof. First let's write the entropy formula of $\tilde{X}$ as in (1.99)

$$H(\tilde{X}) = -\sum_x p(x)\log(p(x)). \tag{1.99}$$

Let N be the number of elements in the range set of $\tilde{X}$. Subtracting $\log N$ from both sides of (1.99), we obtain

$$H(\tilde{X}) - \log N = -\sum_x p(x)\log(p(x)) - \log N. \tag{1.100}$$

Using

$$\sum_x p(x) = 1$$

on the right hand side of (1.100), we get

$$H(\tilde{X}) - \log N = -\sum_x p(x)\log(p(x)) - \log N \sum_x p(x) \tag{1.101}$$

Since $\log N$ is a constant term, when it is carried into the rightmost summation of (1.101), we obtain

$$H(\tilde{X}) - \log N = -\sum_x p(x)\log(p(x)) - \sum_x p(x)\log N. \tag{1.102}$$

In the right hand side of (1.102), the probability mass function $p(x)$ is common term for both summations. Taking the common term out, and combining the logarithmic terms on the right hand side of (1.102), we obtain

$$H(\tilde{X}) - \log N = -\sum_x p(x)\log(Np(x)). \tag{1.103}$$

When $\log(p(x)N)$ in (1.103), is replaced by

$$\frac{\ln(Np(x))}{\ln 2}$$

we obtain

$$H(\tilde{X}) - \log N = \frac{1}{\ln 2}\sum_x p(x)\ln\left(\frac{1}{Np(x)}\right). \tag{1.104}$$

in which for the $\ln(\cdot)$ function, if the property (1.98) is employed, i.e.,

$$\ln\left(\frac{1}{Np(x)}\right) \leq \frac{1}{Np(x)} - 1$$

the expression in (1.104) can be written as

$$H(\tilde{X}) - \log N = \frac{1}{\ln 2}\sum_x p(x)\ln\left(\frac{1}{Np(x)}\right) \leq \frac{1}{\ln 2}\sum_x p(x)\left(\frac{1}{Np(x)} - 1\right) \tag{1.105}$$

which can be written in simpler form as

$$H(\tilde{X}) - \log N \leq \frac{1}{\ln 2}\sum_x p(x)\left(\frac{1}{Np(x)} - 1\right) \tag{1.106}$$

whose right side can be simplified as

$$H(\tilde{X}) - \log N \leq \frac{1}{\ln 2}\left(-\underbrace{\sum_x \frac{1}{N}}_{=\frac{N}{N}}\underbrace{\sum_x p(x)}_{=1}\right). \quad (1.107)$$

Since the right hand side of (1.107) equals to 0, (1.107) gets its final form as

$$H(\tilde{X}) - \log N \leq 0$$

which can be written as

$$H(\tilde{X}) \leq \log N.$$

Example 1.33 Show that

$$I(\tilde{X};\tilde{Y}) \leq \min\{\log|R_{\tilde{X}}|, \log|R_{\tilde{Y}}|\}$$

where $|R_{\tilde{X}}|$ and $|R_{\tilde{Y}}|$ are the number of elements in the range sets of the discrete random variables $\tilde{X}$ and $\tilde{Y}$ respectively.

Solution 1.33 The mutual information between two random variables $\tilde{X}$ and $\tilde{Y}$ is defined as

$$I(\tilde{X};\tilde{Y}) = H(\tilde{X}) - H(\tilde{X}|\tilde{Y}) \quad (1.108)$$

where $H(\tilde{X}|\tilde{Y}) \geq 0$. Then, if we omit $H(\tilde{X}|\tilde{Y})$ from the right hand side of (1.108), we can write

$$I(\tilde{X};\tilde{Y}) \leq H(\tilde{X}). \quad (1.109)$$

Since $H(\tilde{X}) \leq \log|R_{\tilde{X}}|$, then from (1.109) we get

$$I(\tilde{X};\tilde{Y}) \leq \log|R_{\tilde{X}}|. \quad (1.110)$$

In addition, mutual information can also be defined as

$$I(\tilde{X};\tilde{Y}) = H(\tilde{Y}) - H(\tilde{Y}|\tilde{X}) \quad (1.111)$$

and following similar reasoning as in the previous paragraph, we can write

$$I(\tilde{X};\tilde{Y}) \leq \log|R_{\tilde{Y}}|. \tag{1.112}$$

Combining the inequalities (1.110) and (1.112), we obtain

$$I(\tilde{X};\tilde{Y}) \leq \min\{\log|R_{\tilde{X}}|, \log|R_{\tilde{Y}}|\}.$$

Example 1.34 Show that

$$\sum_{x,y} p(x)q(y) = 1 \tag{1.113}$$

where $p(x)$ and $q(y)$ are probability distributions such that

$$\sum_x p(x) = \sum_y q(y) = 1.$$

Solution 1.34 The double summation in (1.113) can be written as

$$\sum_x \sum_y p(x)q(y)$$

which can be evaluated as

$$\underbrace{\sum_x p(x)}_{=1} \underbrace{\sum_y q(y)}_{=1} = 1$$

Example 1.35 Using the log-sum inequality, show that

$$I(\tilde{X};\tilde{Y}) \geq 0.$$

Solution 1.35 According to log-sum inequality, for any two distributions $p(x,y)$ and $q(x,y)$, we have the inequality

$$\sum_{x,y} p(x,y)\log p(x,y) - \sum_{x,y} p(x,y)\log q(x,y) \geq 0$$

where

$$\sum_{x,y} p(x,y) = \sum_{x,y} q(x,y) = 1.$$

The mutual information $I(\tilde{X};\tilde{Y})$ is written in terms of joint and marginal probability mass functions as

$$I(\tilde{X};\tilde{Y}) = \sum_{x,y} p(x,y) \log\left(\frac{p(x,y)}{p(x)p(y)}\right). \tag{1.114}$$

Using the property

$$\log\left(\frac{p(x,y)}{p(x)p(y)}\right) = \log p(x,y) - \log p(x)p(y)$$

of the logarithmic division, the equation in (1.114) can be written as

$$I(\tilde{X};\tilde{Y}) = \sum_{x,y} p(x,y) \log(p(x,y)) - \sum_{x,y} p(x,y) \log(p(x)p(y))$$

where the right hand side is nothing but the log-sum inequality. Hence,

$$I(\tilde{X};\tilde{Y}) \geq 0.$$

Example 1.36 Log-sum inequality is stated as

$$\sum_{x} p(x) \log p(x) \geq \sum_{x} p(x) \log q(x).$$

Verify the log-sum inequality.

Solution 1.36 Consider the difference

$$L = \sum_{x} p(x) \log q(x) - \sum_{x} p(x) \log p(x). \tag{1.115}$$

Combining the logarithmic terms in (1.115), we obtain

$$L = \sum_{x} p(x) \log \frac{q(x)}{p(x)}. \tag{1.116}$$

Using the property

$$\ln x \leq x - 1, \quad x > 0,$$

in (1.116), we obtain

$$L = \sum_x p(x) \log \frac{q(x)}{p(x)} \leq \frac{1}{\ln 2} \sum_x p(x) \left(\frac{q(x)}{p(x)} - 1 \right)$$

where simplifying the right hand side, we get

$$L = \sum_x p(x) \log \frac{q(x)}{p(x)} \leq \frac{1}{\ln 2} \left(\underbrace{\sum_x q(x)}_{=1} - \underbrace{\sum_x p(x)}_{=1} \right). \tag{1.117}$$

Hence, (1.117) reduces to the form

$$L = \sum_x p(x) \log \frac{q(x)}{p(x)} \leq 0$$

which is nothing but the log-sum inequality.

1.5.2 Mutual Information Involving More Than Two Random Variables

The mutual information can also be defined for more than two random variables. For three discrete random variables $\tilde{X}$, $\tilde{Y}$, and $\tilde{Z}$, the mutual information $I(\tilde{X}; \tilde{Y}; \tilde{Z})$ can be calculated using either

$$I(\tilde{X}; \tilde{Y}; \tilde{Z}) = I(\tilde{X}; \tilde{Y}) - I(\tilde{X}; \tilde{Y}|\tilde{Z}) \tag{1.118}$$

or using

$$I(\tilde{X}; \tilde{Y}; \tilde{Z}) = I(\tilde{X}; \tilde{Z}) - I(\tilde{X}; \tilde{Z}|\tilde{Y}) \tag{1.119}$$

or employing

$$I(\tilde{X}; \tilde{Y}; \tilde{Z}) = I(\tilde{Y}; Z) - I(\tilde{Y}; \tilde{Z}|\tilde{X}). \tag{1.120}$$

The relations in Eqs. (1.118), (1.119), and (1.120) can be expressed using Venn diagrams as shown in Fig. 1.21.

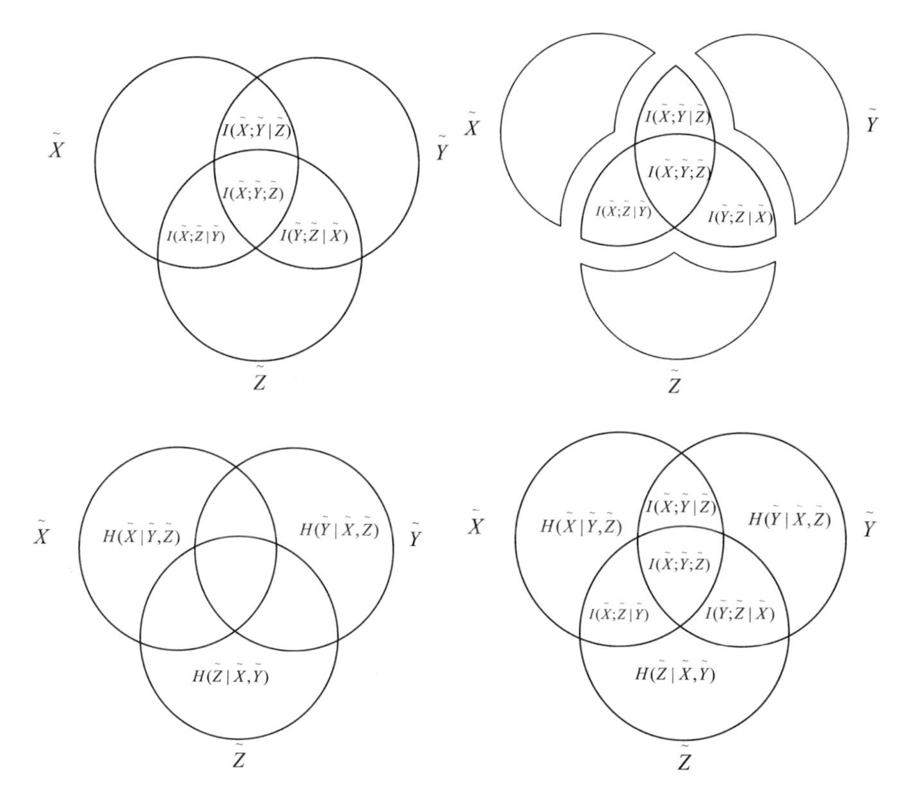

Fig. 1.21 Venn diagram illustration of mutual information

Using (1.118), the joint entropy for three random variables can be expressed using probability mass functions as

$$I(\tilde{X};\tilde{Y};\tilde{Z}) = \sum_{x,y,z} p(x,y,z) \log \frac{p(x,y)p(x,z)p(y,z)}{p(x,y,z)p(x)p(y)p(z)}. \tag{1.121}$$

The mutual information for four random variables $\tilde{W}$, $\tilde{X}$, $\tilde{Y}$, and $\tilde{Z}$ can be calculated using

$$I(\tilde{W};\tilde{X};\tilde{Y};\tilde{Z}) = I(\tilde{W};\tilde{X};\tilde{Y}) - I(\tilde{W};\tilde{X};\tilde{Y}|\tilde{Z}) \tag{1.122}$$

where $I(\tilde{W};\tilde{X};\tilde{Y})$ and $I(\tilde{W};\tilde{X};\tilde{Y}|\tilde{Z})$ can be calculated as

$$I(\tilde{W};\tilde{X};\tilde{Y}) = I(\tilde{W};\tilde{X}) - I(\tilde{W};\tilde{X}|\tilde{Y})$$

and

$$I\left(\tilde{W};\tilde{X};\tilde{Y}|\tilde{Z}\right) = I\left(\tilde{W};\tilde{X}|\tilde{Z}\right) - I\left(\tilde{W};\tilde{X}|\tilde{Y},\tilde{Z}\right)$$

respectively.

1.6 Probabilistic Distance

Let $\tilde{X}_1$ and $\tilde{X}_2$ be two discrete random variables with common range set R. And let $p(x)$ and $q(x)$ be the probability mass functions of these two random variables. The probabilistic distance between these two random variables, or the probabilistic distance between their corresponding distributions is defined as

$$D(\tilde{X}_1||\tilde{X}_2) = \sum_x p(x)\log\frac{p(x)}{q(x)}. \tag{1.123}$$

Example 1.37 For discrete random variables $\tilde{X}_1$ and $\tilde{X}_2$, the common range set, and the distributions on this range set for these two random variables are given as

$$R = \{a, b, c\}$$
$$p(a) = \frac{1}{4} \quad p(b) = \frac{1}{4} \quad p(c) = \frac{1}{2}$$
$$q(a) = \frac{1}{3} \quad q(b) = \frac{1}{6} \quad q(c) = \frac{1}{2}$$

Find the probabilistic distance between $\tilde{X}_1$ and $\tilde{X}_2$, i.e., find the probabilistic distance between the distributions $p(x)$ and $q(x)$.

Solution 1.37 Employing (1.123) for the random variables $\tilde{X}_1$ and $\tilde{X}_2$, we obtain

$$D(\tilde{X}_1||\tilde{X}_2) = \sum_x p(x)\log\frac{p(x)}{q(x)}$$
$$= p(a)\log\frac{p(a)}{q(a)} + p(b)\log\frac{p(b)}{q(b)} + p(c)\log\frac{p(c)}{q(c)}$$

where inserting the given values, we obtain

$$D(\tilde{X}_1||\tilde{X}_2) = \frac{1}{4}\log\frac{1/4}{1/3} + \frac{1}{4}\log\frac{1/4}{1/6} + \frac{1}{2}\log\frac{1/2}{1/2}$$

which is evaluated as

$$D(\tilde{X}_1||\tilde{X}_2) = 0.0425.$$

Example 1.38 Show that, the probabilistic distance $D(\tilde{X}_1||\tilde{X}_2)$ between the random variables $\tilde{X}_1$ and $\tilde{X}_2$ is a nonzero quantity, i.e., $D(\tilde{X}_1||\tilde{X}_2) \geq 0$.

Solution 1.38 The probabilistic distance between two random variables $\tilde{X}_1$ and $\tilde{X}_2$ is defined as

$$D(\tilde{X}_1||\tilde{X}_2) = \sum_x p(x) \log \frac{p(x)}{q(x)}$$

which can be written as

$$D(\tilde{X}_1||\tilde{X}_2) = \sum_x p(x) \log p(x) - \sum_x p(x) \log q(x). \tag{1.124}$$

According to log-sum inequality, we have

$$\sum_x p(x) \log p(x) \geq \sum_x p(x) \log q(x)$$

which implies that $D(\tilde{X}_1||\tilde{X}_2) \geq 0$.

1.7 Jensen's Inequality

Definition A function $g(x)$ is said to be convex over the interval $S = (a, b)$, if for every $x_1, x_2 \in S$ we have

$$g(ax_1 + bx_2) \leq ag(x_1) + bg(x_2)$$

where $a + b = 1$. Convex functions have $\cup$-like shapes, on the other hand, concave functions have $\cap$-like shapes. This is illustrated in Fig. 1.22.

Theorem 1.1 *If $g''(x) > 0$ on the interval $S = (a, b)$, then $g(x)$ is a convex function on the defined interval. On the other hand, if $g''(x) < 0$ on the interval $S = (a, b)$, then $g(x)$ is a convex function on the defined interval.*

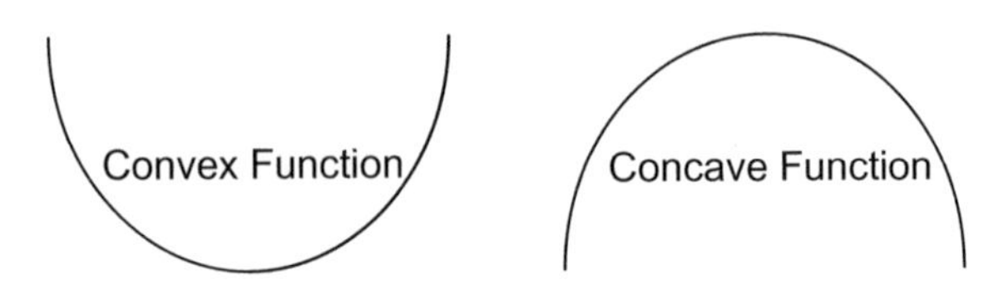

Fig. 1.22 Convex and concave functions

Theorem 1.2 *If $g(x)$ is a convex function and $\tilde{X}$ is a random variable, then*

$$E(g(\tilde{X})) \geq g(E(\tilde{X})) \tag{1.125}$$

where $E(\cdot)$ is the expected value operator.

This theorem is known as Jensen's inequality theorem in the literature. For discrete random variable $\tilde{X}$, the expressions $E(g(\tilde{X}))$ and $(E(\tilde{X}))$ can be calculated as

$$E(g(\tilde{X})) = \sum_x g(x)p(x) \quad E(\tilde{X}) = \sum_x xp(x) \tag{1.126}$$

where $p(x)$ is the probability mass function of the discrete random variable $\tilde{X}$. Using (1.126) *in* (1.125)*, we obtain*

$$\sum_x g(x)p(x) \geq g\left(\sum_x xp(x)\right). \tag{1.127}$$

If $g(x)$ is a concave function, then (1.125) *happens to be as*

$$E(g(\tilde{X})) \leq g(E(\tilde{X})) \tag{1.128}$$

and (1.127) *take the form*

$$\sum_x g(x)p(x) \leq g\left(\sum_x xp(x)\right). \tag{1.129}$$

Example 1.39 The uniformly distributed discrete random variable $\tilde{X}$ has the range set $R_{\tilde{X}} = \{-1, 3\}$. The convex function $g(x)$ is defined as $g(x) = x^2$. Verify the Jensen's inequality theorem.

Solution 1.39 The random variable has the probability mass function

$$p(-1) = \frac{1}{2} \quad p(3) = \frac{1}{2}.$$

According to the Jensen's inequality theorem, we have

$$E\left(g(\tilde{X})\right) \geq g\left(E(\tilde{X})\right)$$

which is written for discrete random variables as

$$\sum_x g(x)p(x) \geq g\left(\sum_x xp(x)\right)$$

where using the given function in the question, we obtain

$$\sum_x x^2 p(x) \geq \left(\sum_x xp(x)\right)^2$$

which is evaluated using the given distribution in the question as

$$(-1)^2\left(\frac{1}{2}\right) + (3)^2\left(\frac{1}{2}\right) \geq \left((-1)\left(\frac{1}{2}\right) + (3)\left(\frac{1}{2}\right)\right)^2. \tag{1.130}$$

When (1.130) is simplified, we obtain

$$5 \geq 1$$

which is a correct expression.

Example 1.40 The uniformly distributed discrete random variable $\tilde{X}$ has the range set $R_{\tilde{X}} = \{x_1, x_2\}$. The convex function $g(x)$ is defined as $g(x) = x^2$. Verify the Jensen's inequality theorem.

Solution 1.40 The random variable has the probability mass function

$$p(x_1) = \frac{1}{2} \quad p(x_2) = \frac{1}{2}.$$

According to the Jensen's inequality theorem, we have

$$E\left(g(\tilde{X})\right) \geq g\left(E(\tilde{X})\right)$$

which is written for discrete random variables as

$$\sum_x g(x)p(x) \geq g\left(\sum_x xp(x)\right)$$

where using the given function in the question, we obtain

$$\sum_x x^2 p(x) \geq \left(\sum_x xp(x)\right)^2$$

which is evaluated using the given distribution in the question as

$$(x_1)^2\left(\frac{1}{2}\right) + (x_2)^2\left(\frac{1}{2}\right) \geq \left((x_1)\left(\frac{1}{2}\right) + (x_2)\left(\frac{1}{2}\right)\right)^2. \tag{1.131}$$

When (1.131) is simplified, we obtain

$$(x_1 - x_2)^2 \geq 0$$

which is a correct inequality.

Remark If $f(x)$ is a concave function, then we have

$$E(g(\tilde{X})) \leq g(E(\tilde{X})).$$

Example 1.41 The logarithmic function is a concave function. The typical graph of the logarithmic function is shown in Fig. 1.23. If we apply the Jensen's inequality.

$$E(g(\tilde{X})) \leq g(E(\tilde{X}))$$

on concave logarithmic function, we obtain

$$E(\log \tilde{X}) \leq \log(E(\tilde{X})). \tag{1.132}$$

Equation (1.132) can be evaluated for discrete random variables as

$$\sum_x \log(x)p(x) \leq \log\left(\sum_x xp(x)\right). \tag{1.133}$$

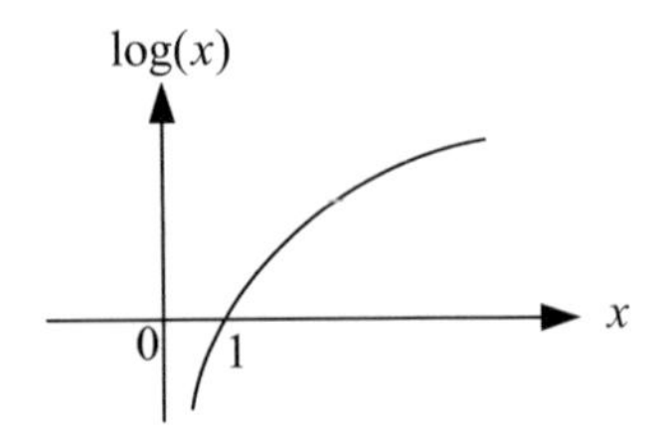

Fig. 1.23 Logarithmic function

Example 1.42 The discrete random variable $\tilde{X}$ has the range set $R_{\tilde{X}} = \{1, 2, 4\}$, and the probability mass function $p(x)$ for the given range set is defined as

$$p(1) = \frac{1}{4} \quad p(2) = \frac{1}{4} \quad p(4) = \frac{1}{2}.$$

Verify (1.133) using the given probability mass function.

Solution 1.42 Expanding (1.133) for the given distribution, we obtain

$$\log(1) \times \frac{1}{4} + \log(2) \times \frac{1}{4} + \log(4) \times \frac{1}{2} \leq \log\left(1 \times \frac{1}{4} + 2 \times \frac{1}{4} + 4 \times \frac{1}{2}\right)$$

which can be simplified as

$$\underbrace{\frac{3}{2}}_{1.25} \leq \underbrace{\log\left(\frac{11}{4}\right)}_{1.46}.$$

Example 1.43 Show that

$$H(\tilde{X}, \tilde{Y}|\tilde{Z}) = H(\tilde{X}|\tilde{Z}) + H(\tilde{Y}|\tilde{X}, \tilde{Z}). \tag{1.134}$$

Solution 1.43 We know that for two discrete random variables $\tilde{A}, \tilde{B}$, the conditional entropy can be written as

$$H(\tilde{A}|\tilde{B}) = H(\tilde{A}, \tilde{B}) - H(\tilde{B}). \tag{1.135}$$

Using (1.135) for (1.134), we get

$$H\left(\underbrace{\tilde{X}, \tilde{Y}}_{\tilde{A}} | \underbrace{\tilde{Z}}_{\tilde{B}}\right) = H\left(\underbrace{\tilde{X}, \tilde{Y}}_{\tilde{A}}, \underbrace{\tilde{Z}}_{\tilde{B}}\right) - H\left(\underbrace{\tilde{Z}}_{\tilde{B}}\right) \tag{1.136}$$

where $H(\tilde{X}, \tilde{Y}, \tilde{Z})$ can be written as

$$H(\tilde{X}, \tilde{Y}, \tilde{Z}) = H(\tilde{Y}|\tilde{X}, \tilde{Z}) + H(\tilde{X}, \tilde{Z}). \tag{1.137}$$

Substituting (1.137) into (1.136), we obtain

$$H(\tilde{X},\tilde{Y}|\tilde{Z}) = H(\tilde{Y}|\tilde{X},\tilde{Z}) + H(\tilde{X}|\tilde{Z}).$$

Following similar steps as in the previous example, it can also be shown that

$$H(\tilde{X},\tilde{Y}|\tilde{Z}) = H(\tilde{X}|\tilde{Y},\tilde{Z}) + H(\tilde{Y}|\tilde{Z}). \tag{1.138}$$

Example 1.44 Show that

$$H(\tilde{X},\tilde{Y},\tilde{Z}) = H(\tilde{X}) + H(\tilde{Y}|\tilde{X}) + H(\tilde{Z}|\tilde{X},\tilde{Y}).$$

Solution 1.44 Using the property

$$H(\tilde{A},\tilde{B}) = H(\tilde{A}|\tilde{B}) + H(\tilde{B})$$

we can write $H(\tilde{X},\tilde{Y},\tilde{Z})$ as

$$H(\tilde{X},\tilde{Y},\tilde{Z}) = H(\tilde{X}) + H(\tilde{Y},\tilde{Z}|\tilde{X})$$

where employing (1.134) in $H(\tilde{Y},\tilde{Z}|\tilde{X})$, we get

$$H(\tilde{X},\tilde{Y},\tilde{Z}) = H(\tilde{X}) + H(\tilde{Y}|\tilde{X}) + H(\tilde{Z}|\tilde{X},\tilde{Y}).$$

Exercise Show that

$$H(\tilde{W},\tilde{X},\tilde{Y},\tilde{Z}) = H(\tilde{W}) + H(\tilde{X}|\tilde{W}) + H(\tilde{Y}|\tilde{W},\tilde{X}) + H(\tilde{Z}|\tilde{W},\tilde{X},\tilde{Y}).$$

Example 1.45 Show that

$$\sum_{x_1,x_2,x_3,x_4,x_5} p(x_1,x_2,x_3,x_4,x_5)\log(x_3|x_2,x_1) = \sum_{x_1,x_2,x_3} p(x_1,x_2,x_3)\log(x_3|x_2,x_1).$$

Solution 1.45 The summation term

$$\sum_{x_1,x_2,x_3,x_4,x_5} p(x_1,x_2,x_3,x_4,x_5)\log(x_3|x_2,x_1)$$

can be expanded as

$$\sum_{\substack{x_1,x_2,x_3\\x_4,x_5}} p(x_1,x_2,x_3,x_4,x_5)\log(x_3|x_2,x_1) = \sum_{x_1,x_2,x_3} \log(x_3|x_2,x_1)\underbrace{\sum_{x_4}\sum_{x_5} p(x_1,x_2,x_3,x_4,x_5)}_{p(x_1,x_2,x_3,x_4)}$$

where the right hand side can be written as

$$\sum_{\substack{x_1,x_2,x_3\\x_4,x_5}} p(x_1,x_2,x_3,x_4,x_5)\log(x_3|x_2,x_1) = \sum_{x_1,x_2,x_3} \log(x_3|x_2,x_1)\underbrace{\sum_{x_4} p(x_1,x_2,x_3,x_4)}_{p(x_1,x_2,x_3)}$$

in which, simplifying the right hand side more, we obtain

$$\sum_{\substack{x_1,x_2,x_3\\x_4,x_5}} p(x_1,x_2,x_3,x_4,x_5)\log(x_3|x_2,x_1) = \sum_{x_1,x_2,x_3} p(x_1,x_2,x_3)\log(x_3|x_2,x_1).$$

We can generalize the obtained result as

$$\begin{aligned}&\sum_{x_1,x_2,\ldots,x_N} p(x_1,x_2,\ldots,x_N)\log(x_i|x_{i-1},\ldots,x_1)\\&= \sum_{x_1,x_2,\ldots,x_i} p(x_1,x_2,\ldots,x_i)\log(x_i|x_{i-1},\ldots,x_1).\end{aligned}$$

Note: The joint probability mass function of N discrete random variables

$$p(x_1,x_2,\ldots,x_N)$$

can be written in terms of the conditional probabilities as

$$\prod_{i=1}^{N} p(x_i|x_{i-1},\ldots,x_1). \tag{1.139}$$

Example 1.46 $p(x_1,x_2,x_3) = p(x_3|x_2,x_1)\times p(x_2|x_1)\times p(x_1)$

Theorem 1.3 *For N discrete random variables* $\tilde{X}_1,\tilde{X}_2,\ldots,\tilde{X}_N$, *the joint entropy satisfies*

$$H(\tilde{X}_1,\tilde{X}_2,\ldots,\tilde{X}_N) = \sum_{i=1}^{N} H(\tilde{X}_i|\tilde{X}_{i-1},\tilde{X}_{i-2},\ldots,\tilde{X}_1).$$

Proof 1.3 The joint entropy of N random variables can be written as

$$H(\tilde{X}_1, \tilde{X}_2, \ldots, \tilde{X}_N) = -\sum_{x_1, x_2, \ldots, x_N} p(x_1, x_2, \ldots, x_N) \log p(x_1, x_2, \ldots, x_N) \quad (1.140)$$

where $p(x_1, x_2, \ldots, x_N)$ is the joint probability mass function of N discrete random variables. Substituting

$$\prod_{i=1}^{N} p(x_i | x_{i-1}, \ldots, x_1)$$

for joint the probability mass function

$$p(x_1, x_2, \ldots, x_N)$$

appearing in the logarithmic expression in (1.140), we obtain

$$H(\tilde{X}_1, \tilde{X}_2, \ldots, \tilde{X}_N) = -\sum_{x_1, x_2, \ldots, x_N} p(x_1, x_2, \ldots, x_N) \log \prod_{i=1}^{N} p(x_i | x_{i-1}, \ldots, x_1)$$

in which substituting

$$\sum_{i=1}^{N} \log p(x_i | x_{i-1}, \ldots, x_1)$$

for

$$\log \prod_{i=1}^{N} p(x_i | x_{i-1}, \ldots, x_1)$$

we get

$$H(\tilde{X}_1, \tilde{X}_2, \ldots, \tilde{X}_N) = -\sum_{x_1, x_2, \ldots, x_N} p(x_1, x_2, \ldots, x_N) \sum_{i=1}^{N} \log p(x_i | x_{i-1}, \ldots, x_1)$$

which can be re-arranged as

$$H(\tilde{X}_1, \tilde{X}_2, \ldots, \tilde{X}_N) = -\sum_{i=1}^{N} \sum_{x_1, x_2, \ldots, x_N} p(x_1, x_2, \ldots, x_N) \log p(x_i | x_{i-1}, \ldots, x_1)$$

where the term

$$\sum_{x_1,x_2,\ldots,x_N} p(x_1,x_2,\ldots,x_N)\log p(x_i|x_{i-1},\ldots,x_1)$$

can be truncated as

$$\sum_{x_1,x_2,\ldots,x_i} p(x_1,x_2,\ldots,x_i)\log p(x_i|x_{i-1},\ldots,x_1).$$

Then, we obtain

$$H(\tilde{X}_1,\tilde{X}_2,\ldots,\tilde{X}_N) = -\sum_{i=1}^{N}\underbrace{\sum_{x_1,x_2,\ldots,x_i} p(x_1,x_2,\ldots,x_i)\log p(x_i|x_{i-1},\ldots,x_1)}_{-\sum_{i=1}^{N} H(\tilde{X}_i|\tilde{X}_{i-1},\ldots,\tilde{X}_1)}$$

which can also be written as

$$H(\tilde{X}_1,\tilde{X}_2,\ldots,\tilde{X}_N) = \sum_{i=1}^{N} H(\tilde{X}_i|\tilde{X}_{i-1},\ldots,\tilde{X}_1).$$

Example 1.47 Show that the conditional entropy is lower additive, i.e.,

$$H(\tilde{X}_1,\tilde{X}_2|\tilde{Y}_1,\tilde{Y}_2) \le H(\tilde{X}_1|\tilde{Y}_1) + H(\tilde{X}_2|\tilde{Y}_2).$$

Solution 1.47 We know that the conditional entropy

$$H(\tilde{X},\tilde{Y}|\tilde{Z})$$

can be written as

$$H(\tilde{X},\tilde{Y}|\tilde{Z}) = H(\tilde{X}|\tilde{Z}) + H(\tilde{Y}|\tilde{X},\tilde{Z}). \tag{1.141}$$

Employing (1.141) for $H(\tilde{X}_1,\tilde{X}_2|\tilde{Y}_1,\tilde{Y}_2)$, we obtain

$$H\left(\tilde{X}_1,\tilde{X}_2|\underbrace{\tilde{Y}_1,\tilde{Y}_2}_{\tilde{Z}}\right) = H\left(\tilde{X}_1|\underbrace{\tilde{Y}_1,\tilde{Y}_2}_{\tilde{Z}}\right) + H\left(\tilde{X}_2|\tilde{X}_1,\underbrace{\tilde{Y}_1,\tilde{Y}_2}_{\tilde{Z}}\right) \tag{1.142}$$

where using the property

$$H(\tilde{X}|\tilde{Y}) \leq H(\tilde{X})$$

we can write

$$H(\tilde{X}_1|\tilde{Y}_1,\tilde{Y}_2) \leq H(\tilde{X}_1|\tilde{Y}_1) \qquad H(\tilde{X}_2|\tilde{X}_1,\tilde{Y}_1,\tilde{Y}_2) \leq H(\tilde{X}_2|\tilde{Y}_2).$$

Thus, (1.142) can be written as

$$H(\tilde{X}_1,\tilde{X}_2|\tilde{Y}_1,\tilde{Y}_2) \leq H(\tilde{X}_1|\tilde{Y}_1) + H(\tilde{X}_2|\tilde{Y}_2).$$

1.8 Fano's Inequality

Assume that we have two discrete correlated random variables $\tilde{X}$ and $\tilde{Y}$, and we want to estimate $\tilde{X}$ using $\tilde{Y}$. Let the estimation of $\tilde{X}$ be $\tilde{Z}$, i.e.,

$$\tilde{Z} = g(\tilde{Y})$$

where g is the estimation function. Then, we can define the probability error random variable as

$$\tilde{E} = \begin{cases} 1 & if\ \tilde{Z} = \tilde{X} \\ 0 & if\ \tilde{Z} \neq \tilde{X} \end{cases}$$

Let's define the probability mass function of $\tilde{E}$ as

$$p_e = Prob(\tilde{E} = 0) \quad q_e = Prob(\tilde{E} = 1) \quad p_e + q_e = 1$$

where p_e indicates the probability of the estimation error.

Fano's inequality states that

$$H(\tilde{X}|\tilde{Y}) \leq 1 + p_e \log(|R_{\tilde{X}}| - 1)$$

where $|R_{\tilde{X}}|$ is the number of elements in the range set of $\tilde{X}$. Now, let's see the proof of Fano's inequality.

Proof The conditional entropy

$$H(\tilde{E},\tilde{X}|\tilde{Y})$$

can be written as

$$H(\tilde{E},\tilde{X}|\tilde{Y}) = H(\tilde{X}|\tilde{Y}) + \underbrace{H(\tilde{E}|\tilde{X},\tilde{Y})}_{=0} \tag{1.143}$$

where $H(\tilde{E}|\tilde{X},\tilde{Y}) = 0$. If both $\tilde{X}$ and $\tilde{Y}$ are known, then $\tilde{E}$ is also know, since $\tilde{X}$ is estimated from $\tilde{Y}$, and $\tilde{E}$ is the estimation error random variable. If $\tilde{E}$ is known, there is no uncertainty about $\tilde{E}$ and this implies that $H(\tilde{E}|\tilde{X},\tilde{Y}) = 0$. Then, from (1.143), we get

$$H(\tilde{E},\tilde{X}|\tilde{Y}) = H(\tilde{X}|\tilde{Y}). \tag{1.144}$$

The conditional entropy

$$H(\tilde{E},\tilde{X}|\tilde{Y})$$

can also be written in an alternative way as

$$H(\tilde{E},\tilde{X}|\tilde{Y}) = H(\tilde{E}|\tilde{Y}) + H(\tilde{X}|\tilde{E},\tilde{Y}). \tag{1.145}$$

Equating the right hand sides of (1.144) and (1.145), we obtain the equality

$$H(\tilde{X}|\tilde{Y}) = H(\tilde{E}|\tilde{Y}) + H(\tilde{X}|\tilde{E},\tilde{Y}) \tag{1.146}$$

where employing the inequality $H(\tilde{E}|\tilde{Y}) \le H(\tilde{E})$, we obtain

$$H(\tilde{X}|\tilde{Y}) \le H(\tilde{E}) + H(\tilde{X}|\tilde{E},\tilde{Y}) \tag{1.147}$$

in which the conditional entropy expression $H(\tilde{X}|\tilde{E},\tilde{Y})$ can be written as

$$H(\tilde{X}|\tilde{E},\tilde{Y}) = \sum_e p(e)H(\tilde{X}|\tilde{E}=e,\tilde{Y}). \tag{1.148}$$

Expanding the right hand side of (1.148) for only e parameter, we obtain

$$H(\tilde{X}|\tilde{E},\tilde{Y}) = \underbrace{p(0)}_{\substack{=Prob(\tilde{E}=0)\\ =p_e}} H(\tilde{X}|\tilde{E}=0,\tilde{Y}) + \underbrace{p(1)}_{\substack{=Prob(\tilde{E}=1)\\ =1-p_e}} H(\tilde{X}|\tilde{E}=1,\tilde{Y}) \tag{1.149}$$

which is written using probability of error mass function as

$$H(\tilde{X}|\tilde{E},\tilde{Y}) = p_e H(\tilde{X}|\tilde{E}=0,\tilde{Y}) + (1-p_e)\underbrace{H(\tilde{X}|\tilde{E}=1,\tilde{Y})}_{=0}. \tag{1.150}$$

In (1.150), we have $H(\tilde{X}|\tilde{E}=1,\tilde{Y})=0$, since $\tilde{E}=1$ means correct estimation is done for $\tilde{X}$ using $\tilde{Y}$. Then, there is no uncertainty about $\tilde{X}$, and we get $H(\tilde{X}|\tilde{E}=1,\tilde{Y})=0$. To put it more clearly, we have

$$H(g(\tilde{Y})|\tilde{Y})=0.$$

In addition, in (1.150), for the expression

$$H(\tilde{X}|\tilde{E}=0,\tilde{Y})$$

we can write

$$H(\tilde{X}|\tilde{E}=0,\tilde{Y})\leq \log(|R_{\tilde{X}}|-1) \tag{1.151}$$

where $|R_{\tilde{X}}|$ is the number of elements in the range set of $\tilde{X}$. Using (1.151) for (1.150), we get

$$H(\tilde{X}|\tilde{E},\tilde{Y})\leq p_e \log(|R_{\tilde{X}}|-1). \tag{1.152}$$

When (1.152) is used for (1.47), we get

$$H(\tilde{X}|\tilde{Y})\leq H(\tilde{E})+p_e \log(|R_{\tilde{X}}|-1). \tag{1.153}$$

For the discrete random variable $\tilde{E}$, we have

$$H(\tilde{E})\leq \log(|R_{\tilde{E}}|)=\log 2\rightarrow H(\tilde{E})\leq 1. \tag{1.154}$$

Using (1.154) for (1.153), we obtain

$$H(\tilde{X}|\tilde{Y})\leq 1+p_e \log(|R_{\tilde{X}}|-1) \tag{1.155}$$

which is nothing but the Fano's inequality. The inequality (1.155) can also be stated as

$$p_e\geq \frac{H(\tilde{X}|\tilde{Y})-1}{\log(|R_{\tilde{X}}|-1)}$$

which indicates a lower bound for the probability of the estimation error.

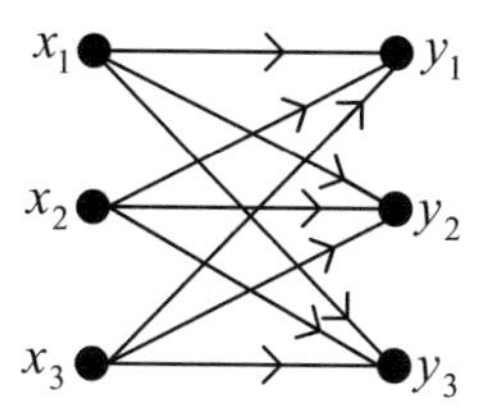

Fig. 1.24 Discrete communication channel for Example 1.48

Example 1.48 For the discrete memoryless channel shown in Fig. 1.24, the input probability mass function $p(x)$, and the channel transition probabilities are given as

$$\begin{aligned}
p(x_1) &= \frac{1}{3} \quad & p(x_2) &= \frac{1}{3} \quad & p(x_3) &= \frac{1}{3} \\
p(y_1|x_1) &= \frac{2}{3} \quad & p(y_2|x_1) &= \frac{1}{6} \quad & p(y_3|x_1) &= \frac{1}{6} \\
p(y_1|x_2) &= \frac{1}{6} \quad & p(y_2|x_2) &= \frac{2}{3} \quad & p(y_3|x_2) &= \frac{1}{6} \\
p(y_1|x_3) &= \frac{1}{6} \quad & p(y_2|x_3) &= \frac{1}{6} \quad & p(y_3|x_3) &= \frac{2}{3}.
\end{aligned}$$

The channel inputs are $x_1 = 0, x_2 = 1, x_3 = 2$ and the channel outputs are $y_1 = 0, y_2 = 1, y_3 = 2$. At the receiver side, we want to estimate the transmitted symbol considering the received symbol. Find the probability of the estimation error.

Solution 1.48 The probability of the estimation error can be calculated using

$$p_e = p(x_1, y_2) + p(x_1, y_3) + p(x_2, y_1) + p(x_2, y_3) + p(x_3, y_1) + p(x_3, y_2)$$

which can be written as

$$\begin{aligned}
p_e = {} & p(y_2|x_1)p(x_1) + p(y_3|x_1)p(x_1) + p(y_1|x_2)p(x_2) + p(y_3|x_2)p(x_2) \\
& + p(y_1|x_3)p(x_3) + p(y_2|x_3)p(x_3)
\end{aligned}$$

leading to

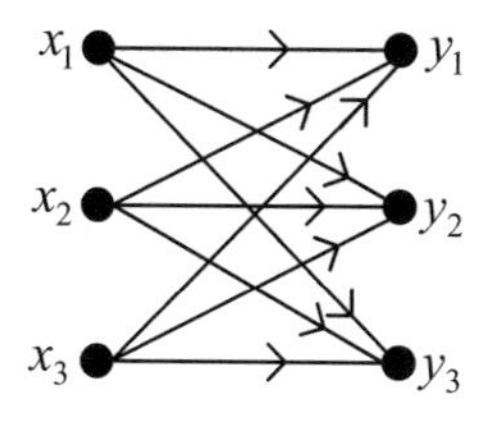

Fig. 1.25 Discrete communication channel for Example 1.49

$$p_e = \frac{1}{6} \times \frac{1}{3} \times 6 \rightarrow p_e = \frac{1}{3}.$$

Example 1.49 For the discrete memoryless channel shown in Fig. 1.25, the input probability mass function $p(x)$, and the channel transition probabilities are given as

$$p(x_1) = \frac{1}{3} \quad p(x_2) = \frac{1}{3} \quad p(x_3) = \frac{1}{3}$$
$$p(y_1|x_1) = \frac{2}{3} \quad p(y_2|x_1) = \frac{1}{6} \quad p(y_3|x_1) = \frac{1}{6}$$
$$p(y_1|x_2) = \frac{1}{6} \quad p(y_2|x_2) = \frac{2}{3} \quad p(y_3|x_2) = \frac{1}{6}$$
$$p(y_1|x_3) = \frac{1}{6} \quad p(y_2|x_3) = \frac{1}{6} \quad p(y_3|x_3) = \frac{2}{3}.$$

The channel inputs are $x_1 = 0, x_2 = 1, x_3 = 2$ and the channel outputs are $y_1 = 0, y_2 = 1, y_3 = 2$. Find the conditional entropy $H(\tilde{X}|\tilde{Y})$.

Solution 1.49 The conditional entropy $H(\tilde{X}|\tilde{Y})$ can be calculated using

$$H(\tilde{X}|\tilde{Y}) = -\sum_{x,y} p(x,y) \log \frac{p(x,y)}{p(y)}$$

where the $p(x,y)$ and $p(y)$ can be calculated as

$$p(x_1, y_1) = p(y_1|x_1)p(x_1) \rightarrow p(x_1, y_1) = \frac{2}{9}$$
$$p(x_1, y_2) = p(y_2|x_1)p(x_1) \rightarrow p(x_1, y_2) = \frac{1}{18}$$
$$p(x_1, y_3) = p(y_3|x_1)p(x_1) \rightarrow p(x_1, y_3) = \frac{1}{18}$$
$$p(x_2, y_1) = p(y_1|x_2)p(x_2) \rightarrow p(x_2, y_1) = \frac{1}{18}$$
$$p(x_2, y_2) = p(y_2|x_2)p(x_2) \rightarrow p(x_2, y_2) = \frac{2}{9}$$
$$p(x_2, y_3) = p(y_3|x_2)p(x_2) \rightarrow p(x_2, y_3) = \frac{1}{18}$$
$$p(x_3, y_1) = p(y_1|x_3)p(x_3) \rightarrow p(x_3, y_1) = \frac{1}{18}$$
$$p(x_3, y_2) = p(y_2|x_3)p(x_3) \rightarrow p(x_3, y_2) = \frac{1}{18}$$
$$p(x_3, y_3) = p(y_3|x_3)p(x_3) \rightarrow p(x_3, y_3) = \frac{2}{9}$$
$$p(y_1) = \frac{1}{3} p(y_2) = \frac{1}{3} p(y_3) = \frac{1}{3}.$$

Then, the conditional entropy can be calculated as

$$H(\tilde{X}|\tilde{Y}) = -\left[3 \times \frac{2}{9}\log\frac{\frac{2}{9}}{\frac{1}{3}} + 6 \times \frac{1}{18}\log\frac{\frac{1}{18}}{\frac{1}{3}}\right] \rightarrow H(\tilde{X}|\tilde{Y}) = 1.25.$$

Example 1.50 Using the results of the previous two examples, verify the Fano's inequality.

Solution 1.50 In our previous two examples, we found that

$$p_e = \frac{1}{3} H(\tilde{X}|\tilde{Y}) = 1.25. \quad (1.156)$$

Fano's inequality states that

$$p_e \geq \frac{H(\tilde{X}|\tilde{Y}) - 1}{\log(|R_{\tilde{X}}| - 1)}$$

in which substituting (1.156), we get

$$\frac{1}{3} \geq \frac{1.25 - 1}{\log(3 - 1)} \rightarrow 0.33 > 0.25\surd$$

which is a correct inequality.

Note: Fano's inequality gives us a lower bound for the estimation error.

Example 1.51 The joint probability mass function of discrete random variables $\tilde{X}$ and $\tilde{Y}$ is given in Table 1.1. Find $H(\tilde{X}|\tilde{Y})$.

Solution 1.51 The conditional entropy $H(\tilde{X}|\tilde{Y})$ can be calculated as

$$H(\tilde{X}|\tilde{Y}) = \sum_y p(y) H(\tilde{X}|y)$$

where $H(\tilde{X}|y)$ is evaluated as

$$H(\tilde{X}|y) = -\sum_x p(x|y) \log p(x|y).$$

Hence, we need the probability mass functions $p(y)$ and $p(x|y)$ for the calculation of the conditional entropy $H(\tilde{X}|\tilde{Y})$. We can find $p(y)$ from $p(x,y)$ using

$$p(y) = \sum_x p(x,y).$$

Table 1.1 Joint probability mass function for Example 1.51

$p(x,y)$

$\tilde{X}$ \ $\tilde{Y}$	a	b	c
d	$\frac{1}{18}$	$\frac{1}{6}$	$\frac{1}{18}$
e	$\frac{1}{6}$	$\frac{1}{18}$	$\frac{1}{18}$
f	$\frac{1}{18}$	$\frac{1}{18}$	$\frac{1}{3}$

Once $p(y)$ is found, $p(x|y)$ can be calculated using

$$p(x|y) = \frac{p(x,y)}{p(y)}.$$

Summing the column probabilities, we obtain probability mass function $p(y)$ as in Table 1.2. Dividing each column by the column sum, we obtain the $p(x|y)$ conditional probability mass function as in Fig. 1.3.

Table 1.2 Joint and marginal probability mass functions

$p(x,y)$

$\tilde{X}$ \ $\tilde{Y}$	a	b	c
d	$\frac{1}{18}$	$\frac{1}{6}$	$\frac{1}{18}$
e	$\frac{1}{6}$	$\frac{1}{18}$	$\frac{1}{18}$
f	$\frac{1}{18}$	$\frac{1}{3}$	$\frac{1}{18}$
$p(y)$	$\frac{5}{18}$	$\frac{10}{18}$	$\frac{3}{18}$

Table 1.3 Conditional probability mass function

$p(x|y)$

$\tilde{X}$ \ $\tilde{Y}$	a	b	c
d	$\frac{1}{5}$	$\frac{3}{10}$	$\frac{1}{3}$
e	$\frac{3}{5}$	$\frac{1}{10}$	$\frac{1}{3}$
f	$\frac{1}{5}$	$\frac{6}{10}$	$\frac{1}{3}$

Using the calculated values in Tables 1.2 and 1.3, we can employ the conditional entropy formula $H(\tilde{X}|\tilde{Y})$ as in

$$H(\tilde{X}|\tilde{Y}) = \sum_y p(y)H(\tilde{X}|y) \rightarrow H(\tilde{X}|\tilde{Y})$$
$$= \frac{5}{18}H\left(\frac{1}{5},\frac{3}{5},\frac{1}{5}\right) + \frac{10}{18}H\left(\frac{3}{10},\frac{1}{10},\frac{1}{10}\right) + \frac{3}{18}H\left(\frac{1}{3},\frac{1}{3},\frac{1}{3}\right)$$

where

$$H\left(\frac{1}{5},\frac{3}{5},\frac{1}{5}\right) = -\left(\frac{1}{5}\log\frac{1}{5} + \frac{3}{5}\log\frac{3}{5} + \frac{1}{5}\log\frac{1}{5}\right) \rightarrow H\left(\frac{1}{5},\frac{3}{5},\frac{1}{5}\right) = 1.3710\,\text{bits/sym}$$
$$H\left(\frac{3}{10},\frac{1}{10},\frac{1}{10}\right) = -\left(\frac{3}{10}\log\frac{3}{10} + \frac{1}{10}\log\frac{1}{10} + \frac{1}{10}\log\frac{1}{10}\right) \rightarrow$$
$$H\left(\frac{3}{10},\frac{1}{10},\frac{1}{10}\right) = 1.1855\,\text{bits/sym}$$
$$H\left(\frac{1}{3},\frac{1}{3},\frac{1}{3}\right) = -\left(\frac{1}{3}\log\frac{1}{3} + \frac{1}{3}\log\frac{1}{3} + \frac{1}{5}\log\frac{1}{3}\right) \rightarrow H\left(\frac{1}{3},\frac{1}{3},\frac{1}{3}\right) = 1.5850\,\text{bits/sym.}$$

Thus, $H(\tilde{X}|\tilde{Y})$ is calculated as

$$H(\tilde{X}|\tilde{Y}) = \frac{5}{18}\times 1.3710 + \frac{10}{18}\times 1.1855$$
$$+ \frac{3}{18}\times 1.5850 \rightarrow H(\tilde{X}|\tilde{Y}) = 1.3036\,\text{bits/sym.}$$

Example 1.52 $\tilde{X}$ and $\tilde{Y}$ are two discrete correlated random variables. Assume that $g(\cdot)$ function is used to estimate discrete random variable $\tilde{X}$ using $\tilde{Y}$. The joint probability mass function of discrete random variables $\tilde{X}$ and $\tilde{Y}$ is given in the Table 1.4. Find the probability of estimation error.

Table 1.4 Joint probability mass function for Example 1.52

$p(x,y)$

$\tilde{X}$ \ $\tilde{Y}$	a	b	c
d	$\frac{1}{18}$	$\frac{1}{6}$	$\frac{1}{18}$
e	$\frac{1}{6}$	$\frac{1}{18}$	$\frac{1}{18}$
f	$\frac{1}{18}$	$\frac{1}{18}$	$\frac{1}{3}$

Solution 1.52 Let the estimation of $\tilde{X}$ be $\tilde{Z}$, i.e., $\tilde{Z} = g(\tilde{Y})$. If $\tilde{Z} = \tilde{X}$, then we have correct estimation. Considering the probabilities in $p(x, y)$, we can define the correct estimation as follows

$$g(\tilde{Y}) = \begin{cases} d & \text{if } \tilde{Y} = b \\ e & \text{if } \tilde{Y} = a \\ f & \text{if } \tilde{Y} = c. \end{cases}$$

The probability of the estimation error can be written as

$$\begin{aligned} p_e =& p(x = d, y = a) + p(x = d, y = c) + p(x = e, y = b) \\ & + p(x = e, y = c) + p(x = f, y = a) + p(x = f, y = b) \end{aligned}$$

which can be evaluated as

$$p_e = 6 \times \frac{1}{18} \rightarrow p_e = \frac{1}{3}$$

Example 1.53 Using the results of the previous two examples verify the Fano's inequality.

Solution 1.53 Fano's inequality states that

$$p_e \geq \frac{H(\tilde{X}|\tilde{Y}) - 1}{\log(|R_{\tilde{X}}| - 1)}$$

where substituting the calculated values found in Examples 1.52 and 1.51, we get

$$\underbrace{p_e}_{\frac{1}{3}} \geq \frac{\overbrace{H(\tilde{X}|\tilde{Y})}^{1.3036} - 1}{\log\left(\underbrace{|R_{\tilde{X}}|}_{3} - 1\right)} \rightarrow \frac{1}{3}$$

$$\geq \frac{0.3036}{\log 2} \rightarrow \frac{1}{3} \geq \frac{0.3036}{\log 2} \rightarrow 0.3333 \geq 0.3036\surd$$

1.9 Conditional Mutual Information

Definition The conditional mutual information between the discrete random variables $\tilde{X}$ and $\tilde{Y}$ given $\tilde{Z}$ is calculated as:

$$\begin{aligned} I(\tilde{X};\tilde{Y}|\tilde{Z}) &= E\left\{\log\frac{p(\tilde{X},\tilde{Y}|\tilde{Z})}{(p(\tilde{X}|\tilde{Z})p(\tilde{Y}|\tilde{Z})}\right\} \\ &= \sum_{x,y,z} p(x,y,z)\log\frac{p(x,y|z)}{p(x|z)p(y|z)} \end{aligned} \tag{1.157}$$

Note: The joint and conditional joint probability mass functions are defined as

$$p(x,y,z) = \text{Prob}\left(\tilde{X} = x, \tilde{Y} = y, \tilde{Z} = z\right)$$

$$p(x,y|z) = \text{Prob}\left(\tilde{X} = x, \tilde{Y} = y|\tilde{Z} = z\right)$$

and we have

$$p(x,y|z) = \frac{p(x,y,z)}{p(z)} \quad p(x|z) = \frac{p(x,z)}{p(z)} \quad p(y|z) = \frac{p(y,z)}{p(z)}$$

Example 1.54 Show that

$$I(\tilde{X};\tilde{Y}|\tilde{Z}) = H(\tilde{X}|\tilde{Z}) - H(\tilde{X}|\tilde{Y},\tilde{Z}). \tag{1.158}$$

Solution 1.54 We have

$$H(\tilde{X}|\tilde{Z}) = -\sum_{x,z} p(x,z)\log p(x|z) \tag{1.159}$$

$$H(\tilde{X}|\tilde{Y},\tilde{Z}) = -\sum_{x,y,z} p(x,y,z)\log p(x|y,z) \tag{1.160}$$

Substituting (1.159) and (1.160) into (1.158), we obtain

$$I(\tilde{X};\tilde{Y}|\tilde{Z}) = -\sum_{x,z} p(x,z)\log p(x|z) + \sum_{x,y,z} p(x,y,z)\log p(x|y,z)$$

where substituting

$$-\sum_{x,y,z} p(x,y,z)\log p(x|z)$$

for

$$-\sum_{x,z} p(x,z)\log p(x|z)$$

we get

$$I(\tilde{X};\tilde{Y}|\tilde{Z}) = -\sum_{x,y,z} p(x,y,z)\log p(x|z) + \sum_{x,y,z} p(x,y,z)\log p(x|y,z)$$

which can be rearranged as

$$I(\tilde{X};\tilde{Y}|\tilde{Z}) = \sum_{x,y,z} p(x,y,z)\log\frac{p(x|y,z)}{p(x|z)} \tag{1.161}$$

in which the probability expression

$$\frac{p(x|y,z)}{p(x|z)}$$

can be manipulated as

$$\frac{p(x|y,z)}{p(x|z)} \to \frac{p(x,y,z)/p(y,z)}{p(x,z)/p(z)} \to \frac{p(x,y,z)p(z)}{p(x,z)p(y,z)} \to \frac{p(x,y|z)p(z)p(z)}{p(x|z)p(z)p(y|z)p(z)}$$
$$\to \frac{p(x,y|z)}{p(x|z)p(y|z)}$$

Thus, we showed that

$$\frac{p(x|y,z)}{p(x|z)} = \frac{p(x,y|z)}{p(x|z)p(y|z)} \tag{1.162}$$

Using (1.162) in (1.161), we get

$$I(\tilde{X};\tilde{Y}|\tilde{Z}) = \sum_{x,y,z} p(x,y,z) \log \frac{p(x,y|z)}{p(x|z)p(y|z)} \tag{1.163}$$

which is nothing but (1.157), i.e., definition of conditional mutual information.

1.9.1 Properties of Conditional Mutual Information

The conditional mutual information definition between $\tilde{X}$ and $\tilde{Y}$ given $\tilde{Z}$ is restated as

$$\begin{aligned} I(\tilde{X};\tilde{Y}|\tilde{Z}) &= E\left\{\log\left(\frac{p(\tilde{X},\tilde{Y}|\tilde{Z})}{p(\tilde{X}|\tilde{Z})p(\tilde{Y}|\tilde{Z})}\right)\right\} \\ &= \sum_{x,y,z} p(x,y,z) \log \frac{p(x,y|z)}{p(x|z)p(y|z)}. \end{aligned} \tag{1.164}$$

Properties *We have the following properties about the conditional mutual information given in* (1.164).

(P1)

$$(a) I(\tilde{X};\tilde{Y}|\tilde{Z})|= H(\tilde{X}|\tilde{Z}) + H(\tilde{Y}|\tilde{Z}) - H(\tilde{X},\tilde{Y}|\tilde{Z})$$
$$(b) I(\tilde{X};\tilde{Y}|\tilde{Z})|= H(\tilde{X}|\tilde{Z}) - H(\tilde{X}|\tilde{Y},\tilde{Z})$$
$$(c) I(\tilde{X};\tilde{Y}|\tilde{Z})|= H(\tilde{Y}|\tilde{Z}) - H(\tilde{Y}|\tilde{X},\tilde{Z})$$

(P2)

$$I(\tilde{X};\tilde{Y}|\tilde{Z}) > 0$$

and

$$I(\tilde{X};\tilde{Y}|\tilde{Z}) = 0$$

if $\tilde{X}$ and $\tilde{Y}$ are conditionally independent given $\tilde{Z}$.

(P3)

$$I(\tilde{X};\tilde{Y}|\tilde{Z}) = I(\tilde{X};\tilde{Y}) + I(\tilde{X};\tilde{Z}|\tilde{Y})$$

Proofs (P1)

(a) The proof of $I(\tilde{X};\tilde{Y}|\tilde{Z}) = H(\tilde{X}|\tilde{Z}) + H(\tilde{Y}|\tilde{Z}) - H(\tilde{X},\tilde{Y}|\tilde{Z})$ can be performed as in

$$\begin{aligned} I(\tilde{X};\tilde{Y}|\tilde{Z}) &= E\left\{\log\left(\frac{p(\tilde{X},\tilde{Y}|\tilde{Z})}{p(\tilde{X}|\tilde{Z})p(\tilde{Y}|\tilde{Z})}\right)\right\} \\ &= \underbrace{E\{-\log p(\tilde{X}|\tilde{Z})\}}_{H(\tilde{X}|\tilde{Z})} + \underbrace{E\{-\log p(\tilde{Y}|\tilde{Z})\}}_{H(\tilde{Y}|\tilde{Z})} - \underbrace{E\{-\log p(\tilde{X},\tilde{Y}|\tilde{Z})\}}_{H(\tilde{X},\tilde{Y}|\tilde{Z})} \\ &= H(\tilde{X}|\tilde{Z}) + H(\tilde{Y}|\tilde{Z}) - H(\tilde{X},\tilde{Y}|\tilde{Z}) \end{aligned}$$

(b) We can prove the equality $I(\tilde{X};\tilde{Y}|\tilde{Z}) = H(\tilde{X}|\tilde{Z}) - H(\tilde{X}|\tilde{Y},\tilde{Z})$ as follows

$$\begin{aligned} I(\tilde{X};\tilde{Y}|\tilde{Z}) &= E\left\{\log\left(\frac{p(\tilde{X},\tilde{Y}|\tilde{Z})}{p(\tilde{X}|\tilde{Z})p(\tilde{Y}|\tilde{Z})}\right)\right\} \\ &= E\{-\log p(\tilde{X}|\tilde{Z})\} + E\left\{\log\frac{p(\tilde{X},\tilde{Y}|\tilde{Z})}{p(\tilde{Y}|\tilde{Z})}\right\} \end{aligned}$$

where the term

$$\frac{p(\tilde{X},\tilde{Y}|\tilde{Z})}{p(\tilde{Y}|\tilde{Z})}$$

can be manipulated as

$$\frac{p(\tilde{X},\tilde{Y}|\tilde{Z})}{p(\tilde{Y}|\tilde{Z})} = \frac{p(\tilde{X},\tilde{Y},\tilde{Z})/p(\tilde{Z})}{p(\tilde{Y},\tilde{Z})/p(\tilde{Z})} \rightarrow \frac{p(\tilde{X},\tilde{Y},\tilde{Z})}{p(\tilde{Y},\tilde{Z})} \rightarrow \frac{p(\tilde{X}|\tilde{Y},\tilde{Z})p(\tilde{Y},\tilde{Z})}{p(\tilde{Y},\tilde{Z})} \rightarrow p(\tilde{X}|\tilde{Y},\tilde{Z}).$$

Hence, we have

$$\begin{aligned} I(\tilde{X};\tilde{Y}|\tilde{Z}) &= E\{-\log p(\tilde{X}|\tilde{Z})\} - E\{\log p(\tilde{X}|\tilde{Y},\tilde{Z})\} \\ &= H(\tilde{X}|\tilde{Z}) - H(\tilde{X}|\tilde{Y},\tilde{Z}). \end{aligned}$$

(c) In a similar manner,

$$I(\tilde{X};\tilde{Y}|\tilde{Z}) = E\left\{\log\left(\frac{p(\tilde{X},\tilde{Y}|\tilde{Z})}{p(\tilde{X}|\tilde{Z})p(\tilde{Y}|\tilde{Z})}\right)\right\}$$
$$= E\{-\log p(\tilde{Y}|\tilde{Z})\} + E\left\{\log\frac{p(\tilde{X},\tilde{Y}|\tilde{Z})}{p(\tilde{X}|\tilde{Z})}\right\}$$

which can be shown to be equal to

$$I(\tilde{X};\tilde{Y}|\tilde{Z}) = H(\tilde{Y}|\tilde{Z}) - H(\tilde{Y}|\tilde{X},\tilde{Z}).$$

(P2) Since,

$$H(\tilde{X}|\tilde{Y},\tilde{Z}) \leq H(\tilde{X}|\tilde{Z})$$

and

$$I(\tilde{X};\tilde{Y}|\tilde{Z}) = H(\tilde{X}|\tilde{Z}) - H(\tilde{X}|\tilde{Y},\tilde{Z})$$

the property

$$I(\tilde{X};\tilde{Y}|\tilde{Z}) \geq 0$$

follows directly.

(P3) $I(\tilde{X};\tilde{Y}|\tilde{Z})$ can be expressed as

$$I(\tilde{X};\tilde{Y}|\tilde{Z}) = E\left\{\log\frac{p(\tilde{X},\tilde{Y},\tilde{Z})}{p(\tilde{X})p(\tilde{Y},\tilde{Z})}\right\} \quad (1.165)$$

where the term

$$\frac{p(\tilde{X},\tilde{Y},\tilde{Z})}{p(\tilde{X})p(\tilde{Y},\tilde{Z})}$$

can be manipulated as follows

$$\frac{p(\tilde{X},\tilde{Y},\tilde{Z})}{p(\tilde{X})p(\tilde{Y},\tilde{Z})} = \frac{p(\tilde{Z}|\tilde{X},\tilde{Y})p(\tilde{X},\tilde{Y})}{p(\tilde{X})p(\tilde{Z}|\tilde{Y})p(\tilde{Y})} \rightarrow \frac{p(\tilde{X},\tilde{Y})}{p(\tilde{X})p(\tilde{Y})}\frac{p(\tilde{Z}|\tilde{X},\tilde{Y})}{p(\tilde{Z}|\tilde{Y})} \rightarrow \frac{p(\tilde{X},\tilde{Y})}{p(\tilde{X})p(\tilde{Y})}\frac{p(\tilde{Z}|\tilde{X},\tilde{Y})p(\tilde{X}|\tilde{Y})}{p(\tilde{Z}|\tilde{Y})p(\tilde{X}|\tilde{Y})}$$
$$\rightarrow \frac{p(\tilde{X},\tilde{Y})}{p(\tilde{X})p(\tilde{Y})}\frac{\overbrace{p(\tilde{Z}|\tilde{X},\tilde{Y})p(\tilde{X}|\tilde{Y})}^{=p(\tilde{X},\tilde{Z}|\tilde{Y})}}{p(\tilde{Z}|\tilde{Y})p(\tilde{X}|\tilde{Y})} \rightarrow \frac{p(\tilde{X},\tilde{Y})}{p(\tilde{X})p(\tilde{Y})}\frac{p(\tilde{X},\tilde{Z}|\tilde{Y})}{p(\tilde{Z}|\tilde{Y})p(\tilde{X}|\tilde{Y})}.$$

Hence, we obtained

$$\frac{p(\tilde{X},\tilde{Y},\tilde{Z})}{p(\tilde{X})p(\tilde{Y},\tilde{Z})}=\frac{p(\tilde{X},\tilde{Y})}{p(\tilde{X})p(\tilde{Y})}\frac{p(\tilde{X},\tilde{Z}|\tilde{X})}{p(\tilde{Z}|\tilde{Y})p(\tilde{X}|\tilde{Y})}. \tag{1.166}$$

Using (1.166) in (1.165), we get

$$I(\tilde{X};\tilde{Y}|\tilde{Z})=E\left\{\log\frac{p(\tilde{X},\tilde{Y})}{p(\tilde{X})p(\tilde{Y})}\frac{p(\tilde{X},\tilde{Z}|\tilde{Y})}{p(\tilde{Z}|\tilde{Y})p(\tilde{X}|\tilde{Y})}\right\}$$

which can be written as

$$I(\tilde{X};\tilde{Y}|\tilde{Z})=\underbrace{E\left\{\log\frac{p(\tilde{X},\tilde{Y})}{p(\tilde{X})p(\tilde{Y})}\right\}}_{I(\tilde{X};\tilde{Y})}+\underbrace{E\left\{\log\frac{p(\tilde{X},\tilde{Z}|\tilde{Y})}{p(\tilde{Z}|\tilde{Y})p(\tilde{X}|\tilde{Y})}\right\}}_{I(\tilde{X};\tilde{Z}|\tilde{Y})}.$$

Thus, we have

$$I(\tilde{X};\tilde{Y}|\tilde{Z})=I(\tilde{X};\tilde{Y})+I(\tilde{X};\tilde{Z}|\tilde{Y}).$$

Example 1.55 Let $\tilde{X}$ be a discrete random variable. If $\tilde{Y}=g(\tilde{X})$, then show that

$$H(\tilde{Y}|\tilde{X})=0.$$

Solution 1.55 The conditional entropy

$$H(\tilde{Y}|\tilde{X})$$

can be calculated using

$$H(\tilde{Y}|\tilde{X})=-\sum_{x,y}p(x,y)\log p(y|x)$$

where

$$\begin{aligned}p(y|x)&=Prob(\tilde{Y}=y|\tilde{X}=x)\\&=Prob(g(\tilde{X})=y|\tilde{X}=x)\\&=1.\end{aligned}$$

Then, we have

$$H(\tilde{Y}|\tilde{X}) = -\sum_{x,y} p(x,y) \log 1 \rightarrow H(\tilde{Y}|\tilde{X}) = 0.$$

Example 1.56 $\tilde{X}$ is a discrete random variable, and $g(\cdot)$ is a function. Show that

$$H(g(\tilde{X})) \leq H(\tilde{X}).$$

Solution 1.56 Using the result of previous example, we can write that

$$\begin{aligned} H(\tilde{X}) &= H(\tilde{X}) + \underbrace{H(g(\tilde{X})|\tilde{X})}_{=0} \\ &= H(\tilde{X}, g(\tilde{X})) \end{aligned} \tag{1.167}$$

where $H(\tilde{X}, g(\tilde{X}))$ can also be written as

$$H(\tilde{X}, g(\tilde{X})) = H(g(\tilde{X})) + \underbrace{H(\tilde{X}|g(\tilde{X}))}_{\geq 0}. \tag{1.168}$$

Using (1.167) and (1.168), we obtain

$$H(\tilde{X}) = H(g(\tilde{X})) + \underbrace{H(\tilde{X}|g(\tilde{X}))}_{\geq 0}$$

which implies that

$$H(g(\tilde{X})) \leq H(\tilde{X}).$$

1.9.2 Markov Chain

The random variables $\tilde{X}$, $\tilde{Y}$, and $\tilde{Z}$ form a Markov chain, if $\tilde{X}$ and $\tilde{Z}$ are conditionally independent given $\tilde{Y}$. The Markov chain is shown as

$$\tilde{X} \rightarrow \tilde{Y} \rightarrow \tilde{Z}.$$

Note that the conditional independence of $\tilde{X}$ and $\tilde{Z}$ given $\tilde{Y}$ implies that

$$p(x,z|y) = p(x|y)p(z|y).$$

1.9.3 Data Processing Inequality for Mutual Information

If $\tilde{X}$, $\tilde{Y}$, and $\tilde{Z}$ form a Markov chain, i.e., $\tilde{X} \to \tilde{Y} \to \tilde{Z}$, then

$$I(\tilde{X};\tilde{Y}) \geq I(\tilde{X};\tilde{Z}).$$

Proof The mutual information between $\tilde{X}$ and $(\tilde{Y},\tilde{Z})$ can be written as

$$I(\tilde{X};\tilde{Y},\tilde{Z}) = I(\tilde{X};\tilde{Y}) + I(\tilde{X};\tilde{Z}|\tilde{Y}) \tag{1.169}$$

where it can be shown that $I(\tilde{X};\tilde{Z}|\tilde{Y}) = 0$. Let's first verify that $I(\tilde{X};\tilde{Z}|\tilde{Y}) = 0$. Since $\tilde{X}$ and $\tilde{Z}$ are conditionally independent given $\tilde{Y}$, then we have

$$p(\tilde{X},\tilde{Z}|\tilde{Y}) = p(\tilde{X}|\tilde{Y})p(\tilde{Z}|\tilde{Y}). \tag{1.170}$$

When (1.170) is used in $I(\tilde{X};\tilde{Z}|\tilde{Y})$, we get

$$\begin{aligned} I(\tilde{X};\tilde{Z}|\tilde{Y}) &= E\left\{ \log \frac{\underbrace{p(\tilde{X},\tilde{Z}|\tilde{Y})}_{=p(\tilde{X}|\tilde{Y})p(\tilde{Z}|\tilde{Y})}}{p(\tilde{X}|\tilde{Y})p(\tilde{Z}|\tilde{Y})} \right\} \\ &= E\{\log 1\} \to 0. \end{aligned}$$

Then, we can write (1.169) as

$$I(\tilde{X};\tilde{Y},\tilde{Z}) = I(\tilde{X};\tilde{Y}). \tag{1.171}$$

The mutual information expression $I(\tilde{X};\tilde{Y},\tilde{Z})$ can also be written alternatively as

$$I(\tilde{X};\tilde{Y},\tilde{Z}) = I(\tilde{X};\tilde{Z}) + \underbrace{I(\tilde{X};\tilde{Y}|\tilde{Z})}_{\geq 0}$$

which implies that

$$I(\tilde{X};\tilde{Z}) \leq I(\tilde{X};\tilde{Y},\tilde{Z}). \tag{1.172}$$

Combining (1.171) and (1.172), we obtain

$$I(\tilde{X};\tilde{Z}) \leq I(\tilde{X};\tilde{Y})$$

which is nothing but the data processing inequality. Let's now solve a numerical examples illustrating the data processing inequality.

Example 1.57 Consider the cascaded binary erasure channel shown in Fig. 1.26. Find a single binary erasure channel that is equal to cascaded binary erasure channel.

Solution 1.57 The probabilities of the symbols y_1, y_2, z_1, z_2 can be calculated using

$$p(y) = \sum_x p(x,y) \rightarrow p(y) = \sum_x p(y|x)p(x)$$

as in

$$p(y_1) = (1-\gamma)\alpha \quad p(y_2) = (1-\gamma)\alpha$$
$$p(z_1) = (1-\gamma)^2\alpha \quad p(z_2) = (1-\gamma)^2(1-\alpha).$$

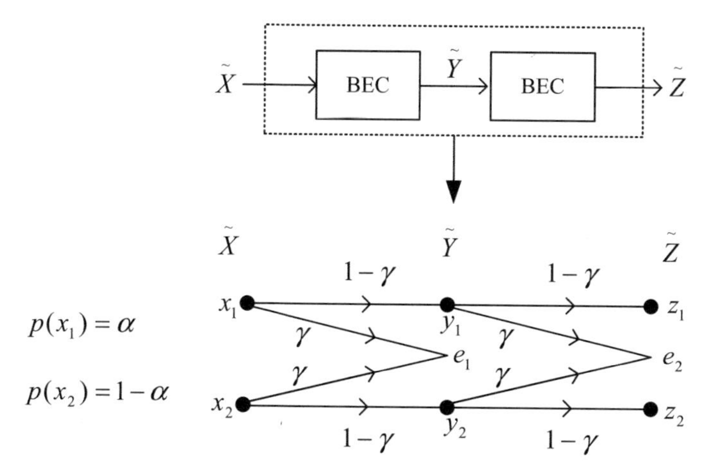

Fig. 1.26 Concatenated binary erasure channel

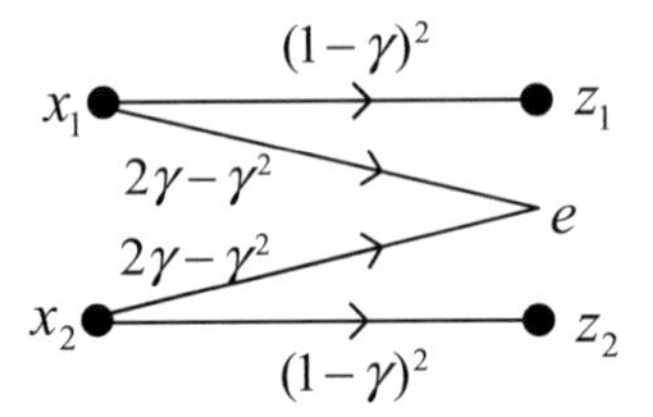

Fig. 1.27 Equivalent binary erasure channel

Considering the probabilities $p(x_1), p(x_2), p(z_1)$, and $p(z_2)$, we can calculate the erasure probability of the equivalent channel as

$$\gamma' = 1 - (1-\gamma)^2 \rightarrow \gamma' = 2\gamma - \gamma^2$$

and an equivalent binary erasure channel of the cascaded channel can be drawn as in Fig. 1.27.

Example 1.58 The mutual information between the input and output of the binary erasure channel depicted in Fig. 1.28 equals to

$$I(\tilde{X}; \tilde{Y}) = (1-\gamma)H_b(\alpha)$$

where

$$H_b(\alpha) = -[\alpha \log \alpha + (1-\alpha)\log(1-\alpha)]$$

such that $H_b(0.5) = 1$. Considering the cascaded binary erasure channel depicted in Fig. 1.29, show that

$$I(\tilde{X}; \tilde{Z}) \leq I(\tilde{X}; \tilde{Y}).$$

Solution 1.58 The mutual information between $\tilde{X}$ and $\tilde{Y}$ can be calculated as

$$I(\tilde{X}; \tilde{Y}) = 1 - \gamma \rightarrow I(\tilde{X}; \tilde{Y}) = 1 - 0.3 \rightarrow I(\tilde{X}; \tilde{Y}) = 0.7.$$

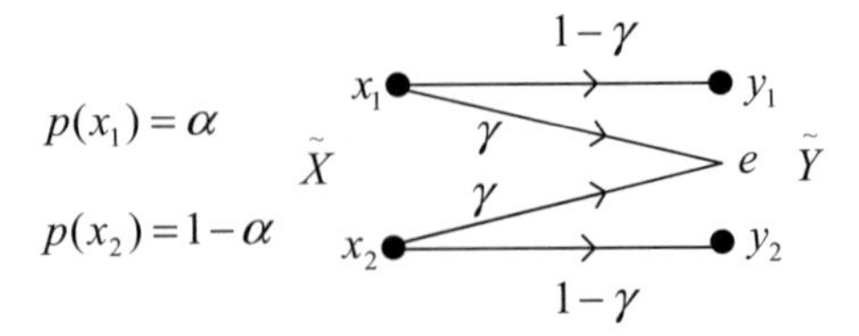

Fig. 1.28 Binary erasure channel for Example 1.58

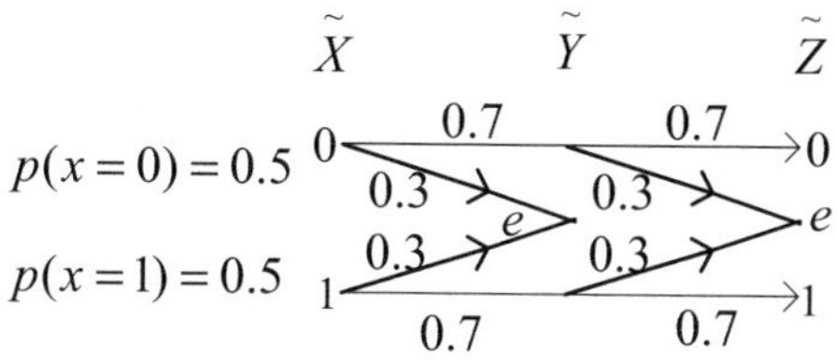

$\tilde{X}$ $\tilde{Y}$ $\tilde{Z}$

$p(x=0)=0.5$ 0 0.7 0.3 e 0.7 0.3 0

$p(x=1)=0.5$ 1 0.3 0.7 0.3 0.7 e 1

Fig. 1.29 Cascaded binary erasure channel

To calculate the mutual information between $\tilde{X}$ and $\tilde{Z}$, we need to find the equivalent model of the cascaded channel. From Example 1.57, the erasure probability of the equivalent channel can be calculated as

$$\gamma' = 2\gamma - \gamma^2 \rightarrow \gamma' = 2 \times 0.3 - 0.7^2 \rightarrow \gamma' = 0.51.$$

The mutual between $\tilde{X}$ and $\tilde{Z}$ can be calculated using $I(\tilde{X};\tilde{Z}) = 1 - \gamma'$ as

$$I(\tilde{X};\tilde{Z}) = 1 - 0.51 \rightarrow I(\tilde{X};\tilde{Z}) = 0.49.$$

Hence, comparing the calculated values $I(\tilde{X};\tilde{Y}) = 0.7$ and $I(\tilde{X};\tilde{Z}) = 0.49$, we see that

$$I(\tilde{X};\tilde{Z}) < I(\tilde{X};\tilde{Y}).$$

Exercise Calculate the erasure probability of single equivalent channel of the cascaded channel consisting of three binary erasure channels.

Example 1.59 Mathematically show that

$$H(\tilde{X}|\tilde{Y}) \leq H(\tilde{X})$$

and

$$H(\tilde{X}|\tilde{Y},\tilde{Z}) \leq H(\tilde{X}|\tilde{Z}).$$

Solution 1.59 The expressions $H(\tilde{X}|\tilde{Y})$ and $H(\tilde{X})$ can be explicitly written as

$$H(\tilde{X}|\tilde{Y}) = -\sum_{x,y} p(x,y) \log p(x|y) \tag{1.173}$$

$$H(\tilde{X}) = -\sum_{x} p(x) \log p(x) \tag{1.174}$$

where $H(\tilde{X})$ can also be written as

$$H(\tilde{X}) = -\sum_{x,y} p(x)\log p(x,y). \tag{1.175}$$

Taking the difference of (1.173) and (1.175), we obtain

$$\begin{aligned} H(\tilde{X}|\tilde{Y}) - H(\tilde{X}) &= -\sum_{x,y} p(x,y)\log p(x|y) + \sum_{x,y} p(x)\log p(x,y) \\ &= \sum_{x,y} p(x,y)\log \frac{p(x)p(y)}{p(x,y)} \end{aligned}$$

in which applying the Jensen's inequality

$$\sum_{x,y} p(x,y)\log q(x,y) \le \log\left(\sum_{x,y} p(x,y)q(x,y)\right)$$

we obtain

$$\begin{aligned} H(\tilde{X}|\tilde{Y}) - H(\tilde{X}) &= \sum_{x,y} p(x,y)\log \underbrace{\frac{p(x)p(y)}{p(x,y)}}_{q(x,y)} \\ &\le \log\left(\sum_{x,y} p(x,y)\underbrace{\frac{p(x)p(y)}{p(x,y)}}_{q(x,y)}\right) \\ &= \log(1) \\ &= 0. \end{aligned}$$

Thus, we get

$$H(\tilde{X}|\tilde{Y}) - H(\tilde{X}) \le 0 \rightarrow H(\tilde{X}|\tilde{Y}) \le H(\tilde{X}).$$

Alternatively,

$$\begin{aligned} H(\tilde{X}) - H(\tilde{X}|\tilde{Y}) &= \sum_{x,y} p(x,y)\log \frac{p(x,y)}{p(x)p(y)} \\ &= D[p(x,y)||p(x)p(y)] \ge 0 \end{aligned}$$

leads to the same result.

To show that

$$H(\tilde{X}|\tilde{Y},\tilde{Z}) \leq H(\tilde{X}|\tilde{Z})$$

let's first write the explicit expression for $H(\tilde{X}|\tilde{Y},\tilde{Z})$ as in

$$H(\tilde{X}|\tilde{Y},\tilde{Z}) = -\sum_{x,y,z} p(x,y,z)\log p(x|y,z). \tag{1.176}$$

The conditional entropy $H(\tilde{X}|\tilde{Z})$ is calculated using

$$H(\tilde{X}|\tilde{Z}) = -\sum_{x,z} p(x,z)\log p(x|z)$$

which can be written as

$$H(\tilde{X}|\tilde{Z}) = -\sum_{x,y,z} p(x,y,z)\log p(x|z). \tag{1.177}$$

Subtracting (1.177) from (1.176), we obtain

$$H(\tilde{X}|\tilde{Y},\tilde{Z}) - H(\tilde{X}|\tilde{Z}) = -\sum_{x,y,z} p(x,y,z)\log p(x|y,z) + \sum_{x,y,z} p(x,y,z)\log p(x|z)$$

which can be written as

$$H(\tilde{X}|\tilde{Y},\tilde{Z}) - H(\tilde{X}|\tilde{Z}) = \sum_{x,y,z} p(x,y,z)\log\frac{p(x|z)}{p(x|y,z)} \tag{1.178}$$

In (1.178), using

$$p(x|y,z) = \frac{p(x,y,z)}{p(y,z)}$$

we obtain

$$H(\tilde{X}|\tilde{Y},\tilde{Z}) - H(\tilde{X}|\tilde{Z}) = \sum_{x,y,z} p(x,y,z)\log\frac{p(x|z)p(y,z)}{p(x,y,z)} \tag{1.179}$$

which can be considered as

$$H(\tilde{X}|\tilde{Y},\tilde{Z}) - H(\tilde{X}|\tilde{Z}) = E\left\{\log\frac{p(\tilde{X}|\tilde{Z})p(\tilde{Y},\tilde{Z})}{p(\tilde{X},\tilde{Y},\tilde{Z})}\right\}. \tag{1.180}$$

Applying Jensen's inequality in (1.179), we obtain

$$\sum_{x,y,z} p(x,y,z)\log\frac{p(x|z)p(y,z)}{p(x,y,z)} \le \log\sum_{x,y,z} p(x,y,z)\frac{p(x|z)p(y,z)}{p(x,y,z)}$$

where the right hand side can be manipulated as

$$\begin{aligned}
&\log\sum_{x,y,z} p(x,y,z)\frac{p(x|z)p(y,z)}{p(x,y,z)} \to \log\sum_{x,y,z} p(x|z)p(y,z)\\
&\to \log\sum_{y,z} p(y,z)\underbrace{\sum_{x} p(x|z)}_{=1}\\
&\to \log\underbrace{\sum_{y,z} p(y,z)}_{=1} \to 0
\end{aligned}$$

Thus, we obtain

$$H(\tilde{X}|\tilde{Y},\tilde{Z}) - H(\tilde{X}|\tilde{Z}) = \sum_{x,y,z} p(x,y,z)\log\frac{p(x|z)p(y,z)}{p(x,y,z)} \le 0$$

i.e.,

$$H(\tilde{X}|\tilde{Y},\tilde{Z}) - H(\tilde{X}|\tilde{Z}) \le 0 \to H(\tilde{X}|\tilde{Y},\tilde{Z}) \le H(\tilde{X}|\tilde{Z})$$

Note: Let's elaborate more on the summation

$$\sum_{y,z} p(y,z)\sum_{x} p(x|z)$$

appearing in the solution. Assume that the random variable $\tilde{Y}, \tilde{Z}$ have two values in their range sets i.e., $R_{\tilde{Y}} = \{y_0, y_1\}$, $R_{\tilde{Z}} = \{z_0, z_1\}$, then

$$\begin{aligned}
\sum_{y,z} p(y,z)\sum_{x} p(x|z) = p(y_0,z_0)\sum_{x} p(x|z_0) + p(y_0,z_1)\sum_{x} p(x|z_1)\\
+ p(y_1,z_0)\sum_{x} p(x|z_0) + p(y_1,z_1)\sum_{x} p(x|z_1)
\end{aligned} \tag{1.181}$$

where we have

$$\sum_{x} p(x|z_i) = 1, \quad i = 0, 1.$$

For instance,

$$\sum_x p(x|z_0) = \sum_x \frac{p(x,z_0)}{p(z_0)} \to \frac{1}{p(z_0)} \underbrace{\sum_x p(x,z_0)}_{p(z_0)} \to 1$$

Then, (1.181) seems to be

$$\begin{aligned}\sum_{y,z} p(y,z) \sum_x p(x|z) = p(y_0,z_0) \underbrace{\sum_x p(x|z_0)}_{=1} + p(y_0,z_1) \underbrace{\sum_x p(x|z_1)}_{=1} \\ + p(y_1,z_0) \underbrace{\sum_x p(x|z_0)}_{=1} + p(y_1,z_1) \underbrace{\sum_x p(x|z_1)}_{=1}\end{aligned} \quad (1.182)$$

Hence, we get

$$\sum_{y,z} p(y,z) \sum_x p(x|z) = p(y_0,z_0) + p(y_0,z_1) + p(y_1,z_0) + p(y_1,z_1) \to 1$$

Second Method:
The explicit expression for $H(\tilde{X}|\tilde{Z}) - H(\tilde{X}|\tilde{Y},\tilde{Z})$ can be written as

$$H(\tilde{X}|\tilde{Z}) - H(\tilde{X}|\tilde{Y},\tilde{Z}) = \sum_{x,y,z} p(x,y,z) \log \frac{p(x,y,z)}{p(x|z)p(y,z)}. \quad (1.183)$$

Referring to (1.183), let $q(x,y,z) = p(x|z)p(y,z)$, and we can show that

$$\sum_{x,y,z} q(x,y,z) = 1.$$

Then, (1.183) is nothing but the probabilistic distance, i.e., relative entropy, between two distributions $p(x,y,z)$ and $q(x,y,z)$, i.e.,

$$H(\tilde{X}|\tilde{Z}) - H(\tilde{X}|\tilde{Y},\tilde{Z}) = D[p(x,y,z), q(x,y,z)]. \quad (1.184)$$

We know that $D[\cdot] \geq 0$, then from (1.184) we have $H(\tilde{X}|\tilde{Z}) - H(\tilde{X}|\tilde{Y},\tilde{Z}) \geq 0$, leading to

$$H(\tilde{X}|\tilde{Y},\tilde{Z}) \leq H(\tilde{X}|\tilde{Z}).$$

Exercise Mathematically show that

$$H(\tilde{X}|\tilde{W},\tilde{Y},\tilde{Z}) \leq H(\tilde{X}|\tilde{Y},\tilde{Z}).$$

Example 1.60 Show that the discrete entropy function $H(\tilde{X})$ is a concave function of the input probabilities, i.e., concave function of $p(x)$.

Solution 1.60 Let's remember the definition of a concave function. A function $g(x)$ is a concave function if

$$g(ax_1 + bx_2) \geq ag(x_1) + bg(x_2), \quad 0 \leq a, b \leq 1, \; a + b = 1 \tag{1.185}$$

is satisfied. Considering (1.185), if $H(\tilde{X})$ is a concave function, then we have

$$H(a\tilde{X}_1 + b\tilde{X}_2) \geq aH(\tilde{X}_1) + bH(\tilde{X}_2). \tag{1.186}$$

Let

$$\tilde{X}_3 = a\tilde{X}_1 + b\tilde{X}_2 \tag{1.187}$$

and $p_3(x)$, $p_2(x)$, and $p_1(x)$ be the probability mass functions of the random variables $\tilde{X}_3$, $\tilde{X}_2$, and $\tilde{X}_1$ respectively. Considering (1.187), we can write that

$$p_3(x) = ap_1(x) + bp_2(x). \tag{1.188}$$

The entropy function of $\tilde{X}_3$ is

$$H(\tilde{X}_3) = -\sum_x p_3(x) \log p_3(x) \tag{1.189}$$

in which substituting (1.188) into (1.189), we obtain

$$H(\tilde{X}_3) = -\sum_x (ap_1(x) + bp_2(x)) \log(ap_1(x) + bp_2(x)). \tag{1.190}$$

The entropy functions of $H(\tilde{X}_1)$ and $H(\tilde{X}_2)$ are given as

$$H(\tilde{X}_1) = -\sum_x p_1(x) \log p_1(x) \quad H(\tilde{X}_2) = -\sum_x p_2(x) \log p_2(x). \tag{1.191}$$

Let's consider the difference

$$H(\tilde{X}_3) - aH(\tilde{X}_1) - bH(\tilde{X}_2) \tag{1.192}$$

where using (1.190) and (1.191), we get

$$-\sum_x (ap_1(x) + bp_2(x)) \log(ap_1(x) + bp_2(x)) + a\sum_x p_1(x) \log p_1(x)$$
$$+ b\sum_x p_2(x) \log p_2(x)$$

which can be simplified as

$$\underbrace{a\sum_x p_1(x) \log\left(\frac{p_1(x)}{ap_1(x) + bp_2(x)}\right)}_{=D[p_1(x)||ap_1(x)+bp_2(x)]\geq 0} + \underbrace{b\sum_x p_2(x) \log\left(\frac{p_2(x)}{ap_1(x) + bp_2(x)}\right)}_{=D[p_2(x)||ap_1(x)+bp_2(x)]\geq 0}. \quad (1.193)$$

Finally, for the difference term in (1.192), we get

$$a \times D[p_1(x)||ap_1(x) + bp_2(x)] + b \times D[p_2(x)||ap_1(x) + bp_2(x)]$$

which is a non-negative quantity. Thus, we can conclude that

$$H(\tilde{X}_3) - aH(\tilde{X}_1) - bH(\tilde{X}_2) \geq 0$$

which is nothing but the criteria for the concavity of $H(\tilde{X})$.

Example 1.61 The random variables $\tilde{X}_1$ and $\tilde{X}_2$ have the range sets

$$R_{\tilde{X}_1} = \{-1, 2\} \quad R_{\tilde{X}_2} = \{-2, 8\}$$

and probability mass functions $p_1(x)$ and $p_2(x)$ explicitly given as

$$p_1(x = -1) = \frac{1}{4} \quad p_1(x = 2) = \frac{3}{4}$$

$$p_2(x = -2) = \frac{7}{8} \quad p_2(x = 8) = \frac{1}{8}.$$

Let

$$\tilde{X}_3 = a\tilde{X}_1 + b\tilde{X}_2$$

where $a = 0.25$, $b = 0.75$. Numerically verify that

$$H(\tilde{X}_3) \geq aH(\tilde{X}_1) + bH(\tilde{X}_2)$$

Solution 1.61 The range set of $\tilde{X}_3$ can be calculated using

$$x_3 = ax_1 + bx_2$$

where $a = 0.25$, $b = 0.75$ and x_1 is chosen from the set $R_{\tilde{X}_1} = \{-1, 2\}$ and x_2 is selected from the range set $R_{\tilde{X}_2} = \{-2, 8\}$, for instance choosing $x_1 = -1$, $x_2 = -2$, we get

$$x_3 = 0.25 \times (-1) + 0.75 \times (-2) \rightarrow x_3 = -1.75.$$

Considering all the x_1, x_2 values, we find the range set of $\tilde{X}_3$ as

$$R_{\tilde{X}_3} = \{-1.75, 5.75, -1, 6.5\}$$

Probability mass function of $\tilde{X}_3$ can be calculated

$$p_3(x_3) = ap_1(x_1) + bp_2(x_2)$$

which results in the following probabilities

$$p_3(x = -1.75) = 0.72 \quad p_3(x = 5.75) = 0.16$$

$$p_3(x = -1) = 0.84 \quad p_3(x = 6.5) = 0.28$$

Using the probability mass functions, we can calculate the entropies as

$$H(\tilde{X}_1) = 0.81 \quad H(\tilde{X}_2) = 0.54 \quad H(\tilde{X}_3) = 1.49$$

$$H(\tilde{X}_3) \geq 0.25H(\tilde{X}_1) + 0.75H(\tilde{X}_2) \rightarrow 1.49 \geq 0.25 \times 0.81 + 0.75 \times 0.54 \rightarrow 1.49 \geq 0.61\surd$$

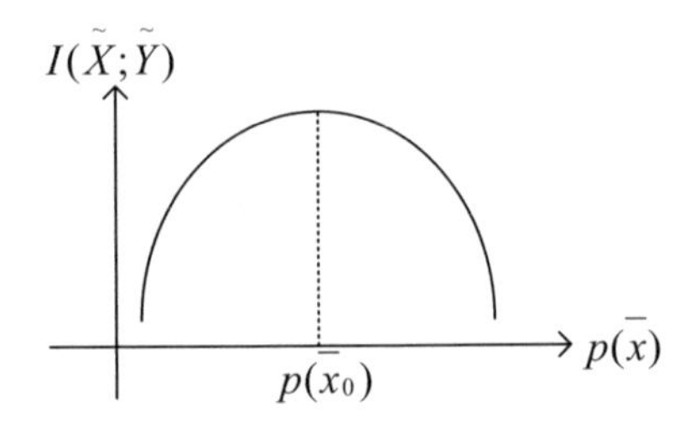

Fig. 1.30 Concavity of mutual information

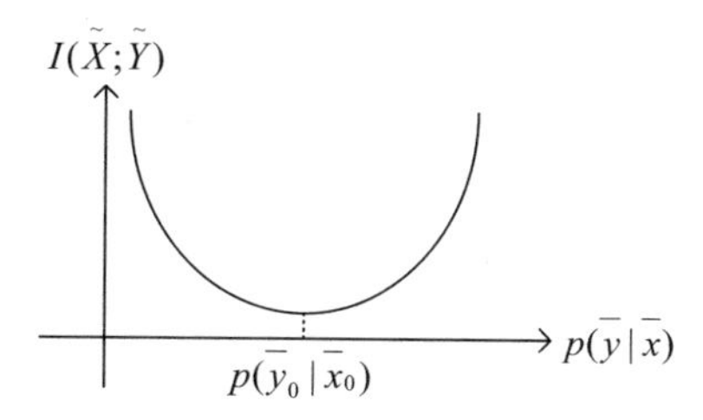

Fig. 1.31 Convexity of mutual information

1.10 Some Properties for Mutual Information

Theorem 1.4 *For known channel transition probabilities, i.e., for fixed* $p(y|x)$*, the mutual information* $I(\tilde{X};\tilde{Y})$ *between channel input and output random variables* $\tilde{X}$ *and* $\tilde{Y}$ *is a concave function of probability mass function of* $\tilde{X}$*, i.e.,* $p(x)$*.*

That is, when the channel is known, i.e., $p(y|x)$ *is fixed, then* $I(\tilde{X};\tilde{Y})$ *is a concave function of* $p(x)$*. Concavity of* $I(\tilde{X};\tilde{Y})$ *is illustrated in Fig.* 1.30.

Theorem 1.5 *The mutual information* $I(\tilde{X};\tilde{Y})$ *between random variables* $\tilde{X}$ *and* $\tilde{Y}$ *is a convex function of* $p(y|x)$ *when* $p(x)$ *is fixed and* $p(y|x)$ *is variable. This situation is illustrated in Fig.* 1.31.

Now let's see the proof of the theorems.

Proof 1.4 Let's first prove the concavity property of the mutual information. If mutual information is a concave function of the input distributions, then it should satisfy

$$I(a\tilde{X}_1 + b\tilde{X}_2;\tilde{Y}) \geq aI(\tilde{X}_1;\tilde{Y}) + bI(\tilde{X}_2;\tilde{Y}) \tag{1.194}$$

where $I(a\tilde{X}_1 + b\tilde{X}_2;\tilde{Y})$ can be written as

$$I(a\tilde{X}_1 + b\tilde{X}_2;\tilde{Y}) = H(a\tilde{X}_1 + b\tilde{X}_2) - H(a\tilde{X}_1 + b\tilde{X}_2|\tilde{Y}). \tag{1.195}$$

In our previous examples, we showed that $H(\cdot)$ is a concave function, i.e.,

$$H(a\tilde{X}_1 + b\tilde{X}_2) \geq aH(\tilde{X}_1) + bH(\tilde{X}_2). \tag{1.196}$$

Employing (1.196) in (1.195), we get

$$I(a\tilde{X}_1 + b\tilde{X}_2;\tilde{Y}) \geq aH(\tilde{X}_1) + bH(\tilde{X}_2) - aH(\tilde{X}_1|\tilde{Y}) - bH(\tilde{X}_2|\tilde{Y})$$

which can be rearranged as

$$I\left(a\tilde{X}_1 + b\tilde{X}_2; \tilde{Y}\right) \geq a\left(\underbrace{H\left(\tilde{X}_1\right) - H\left(\tilde{X}_1|\tilde{Y}\right)}_{I(\tilde{X}_1;\tilde{Y})}\right) + b\left(\underbrace{H\left(\tilde{X}_2\right) - H\left(\tilde{X}_2|\tilde{Y}\right)}_{I(\tilde{X}_2;\tilde{Y})}\right)$$

Hence, we obtained

$$I\left(a\tilde{X}_1 + b\tilde{X}_2; \tilde{Y}\right) \geq aI(\tilde{X}_1; \tilde{Y}) + bI(\tilde{X}_2; \tilde{Y})$$

which is nothing but the criteria for the concavity of $I\left(\tilde{X}; \tilde{Y}\right)$.

Proof 1.5 The mutual information $I\left(\tilde{X}; \tilde{Y}\right)$ between random variables $\tilde{X}$ and $\tilde{Y}$ can be expressed as

$$I\left(\tilde{X}; \tilde{Y}\right) = \sum_{x,y} p(x)p(y|x) \log \frac{p(y|x)}{p(x)}. \tag{1.197}$$

In our previous poof, $p(x)$ was variable, $p(y|x)$ was fixed, and (1.197) gives a non-negative concave function, i.e., $kp \log \frac{k}{p}$ is a concave function. Now, if $p(x)$ is fixed, $p(y|x)$ is variable, it is obvious that (1.197) gives a non-negative convex function.

Example 1.62 For the discrete memoryless channel shown in Fig. 1.32, calculate $I\left(\tilde{X}; \tilde{Y}\right)$ between channel input and output, and show that $I\left(\tilde{X}; \tilde{Y}\right)$ is a concave function of channel input distribution $p(x)$.

Solution 1.62 The mutual information between input and output random variables of the binary erasure channel with erasure probability γ equals to

$$I\left(\tilde{X}; \tilde{Y}\right) = (1 - \gamma)H_b(\alpha) \tag{1.198}$$

where

$$H_b(\alpha) = -[\alpha \log \alpha + (1 - \alpha) \log(1 - \alpha)]$$

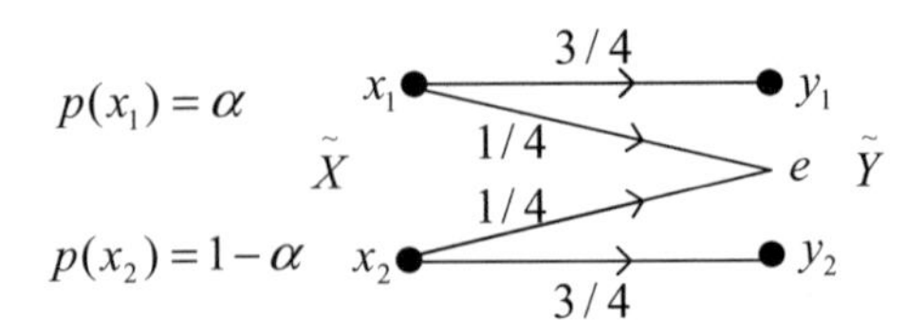

Fig. 1.32 Binary erasure channel for Example 1.62

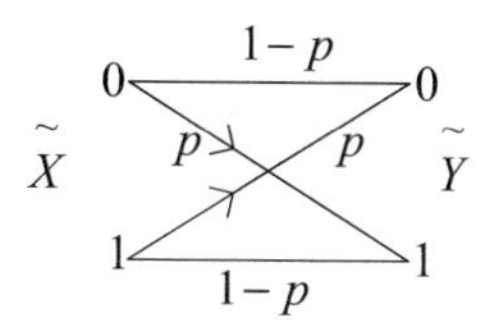

Fig. 1.33 Binary symmetric channel

is a concave function of α. Since γ is a constant, then (1.198) is also a concave function of α.

Example 1.63 For the binary symmetric channel depicted in Fig. 1.33, input random variable has uniform distribution. Show that $I(\tilde{X};\tilde{Y})$ is a convex function of transition probabilities $p(y|x)$, i.e., convex function of p.

Solution 1.63 The mutual information between channel input random variable $\tilde{X}$, and channel output random variable $\tilde{Y}$ can be calculated as

$$I(\tilde{X};\tilde{Y}) = 1 - H_b(p)$$

where $H_b(p)$ is a concave function with peak value '1'. If $H_b(p)$ is a concave function with peak value '1', then $1 - H_b(p)$ is a convex function with minimum value '0'.

Problems

(1) The discrete distributions P and Q are given as

$$P = \left\{ \underbrace{\frac{1}{4}}_{p_1}, \underbrace{\frac{2}{4}}_{p_2}, \underbrace{\frac{1}{4}}_{p_3} \right\} \quad Q = \left\{ \underbrace{\frac{1}{8}}_{q_1}, \underbrace{\frac{1}{8}}_{q_2}, \underbrace{\frac{6}{8}}_{q_3} \right\}.$$

(a) Calculate the probabilistic distance between P and Q, i.e., $D(P||Q) = ?$
(b) Calculate the probabilistic distance between Q and P, i.e., $D(Q||P) = ?$

(2) Consider the binary symmetric channel shown in Fig. 1.P2.

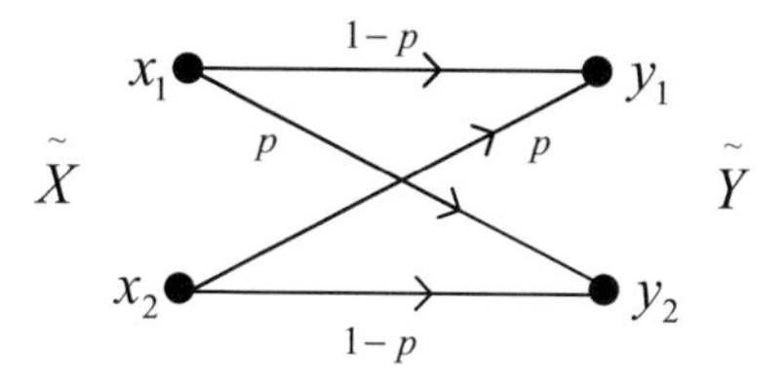

Fig. 1.P2 Binary symmetric channel.

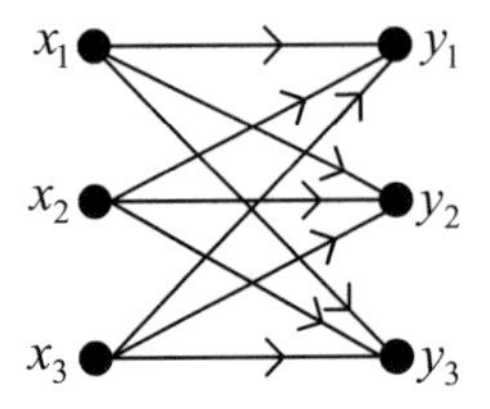

Fig. 1.P3 Discrete memoryless channel

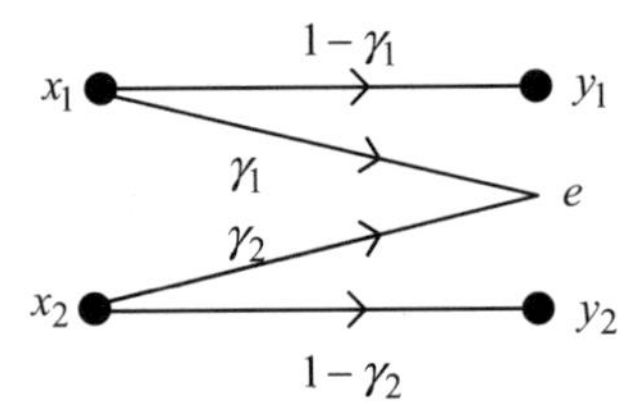

Fig. 1.P4 Binary non-symmetric erasure channel

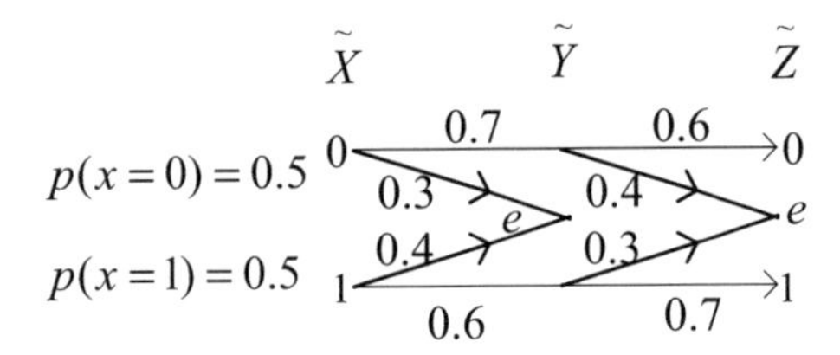

Fig. 1.P5 Cascaded binary erasure channel

Without mathematically calculating the mutual information, sort the mutual information values from largest to smallest, for $p = 1/3, p = 1/4$, and $p = 1/8$.

(3) For the discrete memoryless channel shown in Fig. 1.P3, the input probability mass function $p(x)$, and the channel transition probabilities are given as

$$p(x_1) = \frac{1}{4} \quad p(x_2) = \frac{2}{4} \quad p(x_3) = \frac{1}{4}$$

$$p(y_1|x_1) = \frac{3}{4} \quad p(y_2|x_1) = \frac{1}{8} \quad p(y_3|x_1) = \frac{1}{8}$$

$$p(y_1|x_2) = \frac{1}{8} \quad p(y_2|x_2) = \frac{3}{4} \quad p(y_3|x_2) = \frac{1}{8}$$

$$p(y_1|x_3) = \frac{1}{8} \quad p(y_2|x_3) = \frac{1}{8} \quad p(y_3|x_3) = \frac{3}{4}.$$

The channel inputs are $x_1 = 0, x_2 = 1, x_3 = 2$, and the channel outputs are $y_1 = 0, y_2 = 1, y_3 = 2$.

(a) Calculate $H(\tilde{X}), H(\tilde{Y}), H(\tilde{X}|\tilde{Y}), H(\tilde{Y}|\tilde{X})$.
(b) Calculate the mutual information $I(\tilde{X}|\tilde{Y})$.
(c) At the receiver side, we want to estimate the transmitted symbol considering the received symbol. Find the probability of the estimation error.
(d) Considering the result of part-c, verify the Fano's inequality.

(4) Find the mutual information between input and output of the binary erasure channel given in Fig. 1.P4, assume that $p(x_1) = \alpha$ and $p(x_2) = 1 - \alpha$.
(5) Considering the cascaded binary erasure channel depicted in Fig. 1.P5,

show that

$$I(\tilde{X};\tilde{Z}) \leq I(\tilde{X};\tilde{Y}).$$

Chapter 2
Entropy for Continuous Random Variables Discrete Channel Capacity, Continuous Channel Capacity

In this chapter, we will study the entropy concept for continuous random variables. The capacity of both continuous and discrete channels will be inspected in details. The capacity formula for additive white Gaussian channel is to be derived and the factors that affects the capacity of the additive white Gaussian channel will be elaborated.

2.1 Entropy for Continuous Random Variable

In Chap. 1, we introduced the entropy for discrete random variables. The entropy can also be defined for continuous random variables. The entropy of the continuous random variables is called the differential entropy. Differential entropy has some different properties than the discrete entropy. We will discuss these difference through the chapter.

2.1.1 *Differential Entropy*

Let $\tilde{X}$ be a continuous random variable. The differential entropy $h(\tilde{X})$ for this random variable is defined as

$$h(\tilde{X}) = -\int_{R_{\tilde{X}}} f(x)\log f(x)dx \tag{2.1}$$

where $f(x)$ is the probability density function of the continuous random variable $f(x)$ and $R_{\tilde{X}}$ is the range set of this random variable.

O. Gazi, *Information Theory for Electrical Engineers*, Signals and Communication Technology, https://doi.org/10.1007/978-981-10-8432-4_2

Note: Capital letter '*H*' is used to denote the entropy of discrete random variables, whereas small letter '*h*' is used to denote the entropy of the continuous random variables.

Example 2.1 $\tilde{X}$ is a continuous random variable uniformly distributed on the interval $[0\ a]$. Calculate the differential entropy for this random variable.

Solution 2.1 The probability density function of the given continuous random variable is

$$f(x) = \frac{1}{a} \quad 0 \leq x \leq a$$

and the range set of this random variable is $R_{\tilde{X}} = [0\ a]$. Applying the differential entropy formula

$$h(\tilde{X}) = -\int_{R_{\tilde{X}}} f(x) \log f(x) dx \tag{2.2}$$

on the given distribution, we can calculate the differential entropy of the given random variable as

$$h(\tilde{X}) = -\int_0^a \frac{1}{a} \log \frac{1}{a} dx \rightarrow h(\tilde{X}) = \log a.$$

From the obtained result, we see that for $0 < a < 1$, $\log a < 0$. This means that we can have negative differential entropy. Hence, unlike discrete entropy the differential entropy can be negative.

Example 2.2 $\tilde{X}$ is a continuous random variable with zero mean normal distribution, i.e.,

$$\tilde{X} \sim N(0, \sigma^2).$$

Calculate the differential entropy of this random variable.

Solution 2.2 The range set for this random variable is $R_{\tilde{X}} = (-\infty\ \infty)$, and the probability density function of the given random variable is

$$f(x) = \frac{1}{\sigma\sqrt{2\pi}} e^{-\frac{x^2}{2\sigma^2}}. \tag{2.3}$$

Let's use (2.3) only in the logarithmic part of (2.2) as follows

$$h(\tilde{X}) = -\int_{\infty}^{\infty} f(x) \log \frac{1}{\sigma\sqrt{2\pi}} e^{-\frac{x^2}{2\sigma^2}} dx. \tag{2.4}$$

The logarithmic expression in (2.4) can be simplified as

$$\begin{aligned}
\log \frac{1}{\sigma\sqrt{2\pi}} e^{-\frac{x^2}{2\sigma^2}} &= -\log\sqrt{2\pi\sigma^2} + \log e^{-\frac{x^2}{2\sigma^2}} \\
&= -\log\sqrt{2\pi\sigma^2} + \frac{1}{\ln 2} \ln e^{-\frac{x^2}{2\sigma^2}} \\
&= -\frac{1}{\ln 2} \times \ln\sqrt{2\pi\sigma^2} - \frac{1}{\ln 2} \times \frac{x^2}{2\sigma^2} \\
&= -\frac{1}{\ln 2} \times \left(\ln\sqrt{2\pi\sigma^2} + \frac{x^2}{2\sigma^2}\right).
\end{aligned}$$

Thus, we have

$$\log \frac{1}{\sigma\sqrt{2\pi}} e^{-\frac{x^2}{2\sigma^2}} = -\frac{1}{\ln 2} \times \left(\ln\sqrt{2\pi\sigma^2} + \frac{x^2}{2\sigma^2}\right). \tag{2.5}$$

When the simplified expression (2.5) is substituted into (2.4), we obtain

$$h(\tilde{X}) = \frac{1}{\ln 2} \int_{\infty}^{\infty} f(x)\left(\ln\sqrt{2\pi\sigma^2} + \frac{x^2}{2\sigma^2}\right) dx$$

which can be written as

$$h(\tilde{X}) = \frac{\ln\sqrt{2\pi\sigma^2}}{\ln 2} \underbrace{\int_{\infty}^{\infty} f(x) dx}_{=1} + \frac{1}{2\sigma^2 \ln 2} \underbrace{\int_{\infty}^{\infty} x^2 f(x)}_{\sigma^2}. \tag{2.6}$$

From (2.6), we obtain

$$h(\tilde{X}) = \frac{1}{\ln 2}\left(\ln\sqrt{2\pi\sigma^2} + \frac{1}{2}\right) \tag{2.7}$$

where replacing the constant term '1/2' by its logarithmic expression

$$\frac{1}{2} = \frac{1}{2}\ln \mathrm{e}$$

we obtain

$$h(\tilde{X}) = \frac{1}{\ln 2}\left(\frac{1}{2}\ln 2\pi\sigma^2 + \frac{1}{2}\ln \mathrm{e}\right) \tag{2.8}$$

which can be simplified as

$$h(\tilde{X}) = \frac{1}{2}\log 2\pi\mathrm{e}\sigma^2. \tag{2.9}$$

Exercise If $\tilde{X} \sim N(m, \sigma^2)$ find $h(\tilde{X})$.

Solution If same steps as in the previous example followed, it is seen that $h(\tilde{X})$ does not change, that is $h(\tilde{X})$ happens to be

$$h(\tilde{X}) = \frac{1}{2}\log 2\pi\mathrm{e}\sigma^2.$$

This means that the entropy of a Gaussian random variable in only affected by its variance.
Note: The variance σ^2 of a continuous random variable $\tilde{X}$ is calculated as

$$\begin{aligned}\sigma^2 &= E(\tilde{X}^2) - \left[E(\tilde{X})\right]^2 \\ &= \int_{-\infty}^{\infty} x^2 f(x)dx - \left[\int_{-\infty}^{\infty} xf(x)dx\right]^2\end{aligned}$$

Example 2.3 If $\tilde{X} \sim N(0, 1)$, find $h(\tilde{X})$.

Solution 2.3 The entropy of the normal random variable is calculated as

$$h(\tilde{X}) = \frac{1}{2}\log 2\pi\mathrm{e}\sigma^2$$

in which substituting $\sigma^2 = 1$, we obtain

$$h(\tilde{X}) = \frac{1}{2}\log 2\pi\mathrm{e} \rightarrow h(\tilde{X}) = 2.0471 \text{ bits}.$$

Exercise If $\tilde{X} \sim N(m, \sigma^2)$ and $\tilde{Y} = \tilde{X} + a$ where a is a constant, show that $var(\tilde{Y}) = var(\tilde{X})$ which implies that $h(\tilde{X}) = h(\tilde{Y})$.

Exercise If $\tilde{X} \sim N(m, \sigma^2)$ and $\tilde{Y} = 8\tilde{X}$, express $h(\tilde{Y})$ in terms of $h(\tilde{X})$.

2.1.2 Joint and Conditional Entropies for Continuous Random Variables

The entropy of the continuous random variables is called differential entropy in the literature.

Joint Differential Entropy

The differential entropy of N continuous random variables is defined as

$$h(\tilde{X}_1, \tilde{X}_2, \ldots, \tilde{X}_N) = -\int_{R_s} f(x_1, x_2, \ldots, x_N) \log f(x_1, x_2, \ldots, x_N) dx_1 dx_2, \ldots, dx_N \tag{2.10}$$

where $f(x_1, x_2, \ldots, x_N)$ is the joint probability density function of these N continuous random variables, and R_s is the joint range set of the random variables.

Conditional Differential Entropy

Let $\tilde{X}$ and $\tilde{Y}$ be two continuous random variables. The conditional entropy of $\tilde{X}$ given $\tilde{Y}$ is defined as

$$h(\tilde{X}|\tilde{Y}) = \int f(x, y) \log f(x|y) dxdy. \tag{2.11}$$

If

$$f(x|y) = \frac{f(x, y)}{f(y)}$$

is substituted into (2.11), we obtain

$$h(\tilde{X}|\tilde{Y}) = h(\tilde{X}, \tilde{Y}) - h(\tilde{Y}).$$

In a similar manner, starting with

$$h(\tilde{Y}|\tilde{X}) = \int f(x,y)\log f(y|x)dxdy$$

and using

$$f(y|x) = \frac{f(x,y)}{f(x)}$$

we can show that

$$h(\tilde{Y}|\tilde{X}) = h(\tilde{X},\tilde{Y}) - h(\tilde{X}).$$

2.1.3 *The Relative Entropy of Two Continuous Distribution*

Let $f(x)$ and $g(x)$ be two probability density functions of the continuous random variables $\tilde{X}$ and $\tilde{Y}$. The distance, i.e., relative entropy, between $f(x)$ and $g(x)$ is defined as

$$D[f(x)||g(x)] = \int f(x)\log\frac{f(x)}{g(x)}dx. \tag{2.12}$$

Example 2.4 The normal distributions $f(x)$ and $g(x)$ are defined as

$$f(x) = \frac{1}{\sqrt{2\pi\sigma_1^2}}e^{-\frac{x^2}{2\sigma_1^2}} \quad g(x) = \frac{1}{\sqrt{2\pi\sigma_2^2}}e^{-\frac{x^2}{2\sigma_2^2}}$$

Find

$$D[f(x)||g(x)].$$

Solution 2.4 For the given distributions, using (2.12), we obtain

$$D[f(x)||g(x)] = \int_{-\infty}^{\infty}\frac{1}{\sqrt{2\pi\sigma_1^2}}e^{-\frac{x^2}{2\sigma_1^2}}\log\frac{\frac{1}{\sqrt{2\pi\sigma_1^2}}e^{-\frac{x^2}{2\sigma_1^2}}}{\frac{1}{\sqrt{2\pi\sigma_2^2}}e^{-\frac{x^2}{2\sigma_2^2}}}dx$$

which can be written as

$$D[f(x)||g(x)] = \frac{1}{\sqrt{2\pi\sigma_1^2}} \int_{-\infty}^{\infty} e^{-\frac{x^2}{2\sigma_1^2}} \log \frac{\sigma_2}{\sigma_1} e^{-x^2\left(\frac{1}{2\sigma_1^2} - \frac{1}{2\sigma_2^2}\right)} dx \tag{2.13}$$

where expanding the logarithmic term, we get

$$D[f(x)||g(x)] = \frac{1}{\sqrt{2\pi\sigma_1^2}} \int_{-\infty}^{\infty} e^{-\frac{x^2}{2\sigma_1^2}} \left(\log \frac{\sigma_2}{\sigma_1} - \frac{x^2}{\ln 2} \left(\frac{1}{2\sigma_1^2} - \frac{1}{2\sigma_2^2} \right) \right) dx$$

$$= \log \frac{\sigma_2}{\sigma_1} \underbrace{\frac{1}{\sqrt{2\pi\sigma_1^2}} \int_{-\infty}^{\infty} e^{-\frac{x^2}{2\sigma_1^2}} dx}_{=1} - \frac{1}{\ln 2} \left(\frac{1}{2\sigma_1^2} - \frac{1}{2\sigma_2^2} \right) \underbrace{\frac{1}{\sqrt{2\pi\sigma_1^2}} \int_{-\infty}^{\infty} x^2 e^{-\frac{x^2}{2\sigma_1^2}} dx}_{\sigma_1^2}$$

$$= \log \frac{\sigma_2}{\sigma_1} - \frac{1}{\ln 2} \left(\frac{1}{2} - \frac{\sigma_1^2}{2\sigma_2^2} \right).$$

Hence, we have

$$D[f(x)||g(x)] = \log \frac{\sigma_2}{\sigma_1} - \frac{1}{\ln 2} \left(\frac{1}{2} - \frac{\sigma_1^2}{2\sigma_2^2} \right). \tag{2.14}$$

It is clear from (2.14) that, when $\sigma_2 = \sigma_1 = 1$, we have $D[f(x)||g(x)] = 0$.

Exercise If

$$f(x) = \frac{1}{\sqrt{2\pi\sigma_1^2}} e^{-\frac{(x-m_1)^2}{2\sigma_1^2}} \quad g(x) = \frac{1}{\sqrt{2\pi\sigma_2^2}} e^{-\frac{(x-m_2)^2}{2\sigma_2^2}}$$

Find

$$D[f(x)||g(x)].$$

Example 2.5 The distributions $f(x)$ and $g(x)$ are defined on the interval $[0\ a]$ as $f(x) = 1/a$, and $g(x) = Ke^{-bx}$. Find

$$D[f(x)||g(x)] \quad \text{and} \quad D[g(x)||f(x)].$$

Solution 2.5 Using

$$D[f(x)||g(x)] = \int f(x)\log\frac{f(x)}{g(x)}dx$$

for the given intervals, we obtain

$$D[f(x)||g(x)] = \int_0^a \frac{1}{a}\log\frac{\frac{1}{a}}{Ke^{-bx}}dx \rightarrow D[f(x)||g(x)] = \frac{ab}{2\ln 2} - \log Ka.$$

Note: Relative entropy is not commutative, i.e.,

$$D[f(x)||g(x)] \neq D[g(x)||f(x)].$$

Exercise Find the entropy of the exponential distribution $f(x) = \lambda e^{-\lambda}, x \geq 0$.

Exercise Find the distance between the distributions $f(x) = \lambda_1 e^{-\lambda_1 x}, x \geq 0$ and $g(x) = \lambda_2 e^{-\lambda_2 x}, x \geq 0$.

2.2 Mutual Information for Continuous Random Variables

The mutual information $I(\tilde{X}; \tilde{Y})$ between continuous random variables $\tilde{X}$ and $\tilde{Y}$ with joint probability density function $f(x, y)$ is defined as

$$I(\tilde{X}; \tilde{Y}) = \int f(x, y)\log\frac{f(x, y)}{f(x)f(y)}dxdy. \tag{2.15}$$

Using the expressions of differential entropy and conditional differential entropy, $I(\tilde{X}; \tilde{Y})$ can be expressed either using

$$I(\tilde{X}; \tilde{Y}) = h(\tilde{X}) - h(\tilde{X}|\tilde{Y})$$

or using

$$I(\tilde{X}; \tilde{Y}) = h(\tilde{Y}) - h(\tilde{Y}|\tilde{X}).$$

Example 2.6 If $f(x)$ be a continuous distribution function, and $k(x)$ is any function. Show that

$$\int_{-\infty}^{\infty} f(x)\log k(x)dx \leq \log \int_{-\infty}^{\infty} f(x)k(x)dx.$$

Solution 2.6 According to Jensen's inequality, if $\phi(\cdot)$ is a concave function, then we have

$$E\{\phi(k(x))\} \leq \phi\{E(k(x))\}.$$

If the concave function $\phi(\cdot)$ is chosen as $\log(\cdot)$, then we get

$$E\{\log(k(x))\} \leq \log\{E(k(x))\} \tag{2.16}$$

which can be written explicitly as

$$\int_{-\infty}^{\infty} f(x)\log k(x)dx \leq \log \int_{-\infty}^{\infty} f(x)k(x)dx. \tag{2.17}$$

Theorem 2.1 *The relative entropy is a non-negative quantity, i.e.,*

$$D[f(x)||g(x)] \geq 0. \tag{2.18}$$

Equality occurs if $f(x) = g(x)$.

Proof 2.1 The relative entropy is defined as

$$D[f(x)||g(x)] = \int f(x)\log\frac{f(x)}{g(x)}dx.$$

Then

$$-D[f(x)||g(x)]$$

happens to be

$$-D[f(x)||g(x)] = \int f(x)\log\frac{g(x)}{f(x)}dx.$$

in which employing (2.17), we get

$$-D[f(x)||g(x)] = \int f(x)\log\frac{g(x)}{f(x)}dx \le \log\int f(x)\frac{g(x)}{f(x)}dx$$
$$= \underbrace{\log\underbrace{\int g(x)dx}_{=1}}_{=0}.$$

That is,

$$-D[f(x)||g(x)] \le 0$$

which means that

$$D[f(x)||g(x)] \ge 0.$$

Example 2.7 Write the mutual information expression for the continuous random variables in terms of the relative entropy operator $D[\cdot]$.

Solution 2.7 The differential entropy between two continuous random variables is defined as

$$I(\tilde{X};\tilde{Y}) = \int f(x,y)\log\frac{f(x,y)}{f(x)f(y)}dxdy$$

which can be written in terms of the relative entropy as

$$I(\tilde{X};\tilde{Y}) = D[f(x,y)||f(x)f(y)].$$

Using the property $D[\cdot] \ge 0$, we can draw the following corollaries.

Corollary 1 *$I(\tilde{X};\tilde{Y}) \ge 0$, since $I(\tilde{X};\tilde{Y}) = D[f(x,y)||f(x)f(y)]$.*

Corollary 2 *$h(\tilde{X}|\tilde{Y}) < h(\tilde{X})$, since $I(\tilde{X};\tilde{Y}) = h(\tilde{X}) - h(\tilde{X}|\tilde{Y})$, and $I(\tilde{X};\tilde{Y}) \ge 0$.*

Theorem 2.2 *The joint differential entropy of N continuous random variables satisfy*

$$h(\tilde{X}_1,\tilde{X}_2,\ldots,\tilde{X}_N) \le \sum_{i=1}^{N} h(\tilde{X}_i|\tilde{X}_1,\tilde{X}_2,\ldots,\tilde{X}_{i-1}). \tag{2.19}$$

Theorem 2.3 *Differential entropy is invariant to the shifting operation, i.e.,*

$$h(\tilde{X}+c) = h(\tilde{X}). \tag{2.20}$$

Theorem 2.4 *For random variables scaling, differential entropy satisfies*

$$h(a\tilde{X}) = h(\tilde{X}) + \log|a|. \tag{2.21}$$

Proof 2.4 We will prove only Theorem 2.4. The proofs of the other theorems are similar to those proofs of the discrete entropy having similar theorems in Chap. 1. Let

$$\tilde{Y} = a\tilde{X}$$

and assume that random variable $\tilde{X}$ is defined on $(-\infty\ \infty)$. Then, we have

$$h(\tilde{Y}) = -\int_{-\infty}^{\infty} f(y)\log f(y)dy. \tag{2.22}$$

Since $\tilde{Y} = a\tilde{X}$ then, from probability course, we have

$$f_{\tilde{Y}}(y) = \frac{1}{|a|}f_{\tilde{X}}\left(\frac{y}{a}\right) \tag{2.23}$$

where $f_{\tilde{Y}}(\cdot)$ is the probability density function of $\tilde{Y}$, and $f_{\tilde{X}}(\cdot)$ is the probability density function of the random variable $\tilde{X}$. Using (2.23) in (2.22), we obtain

$$h(\tilde{Y}) = -\int_{-\infty}^{\infty} \frac{1}{|a|}f_{\tilde{X}}\left(\frac{y}{a}\right)\log\left(\frac{1}{|a|}f_{\tilde{X}}\left(\frac{y}{a}\right)\right)dy. \tag{2.24}$$

Since $\tilde{Y} = a\tilde{X}$, then we have $y = ax$. If $a > 0$, then $dy = adx$, this means

$$h(\tilde{Y}) = -\int_{-\infty}^{\infty} \frac{1}{a}f_{\tilde{X}}\left(\frac{y}{a}\right)\log\left(\frac{1}{a}f_{\tilde{X}}\left(\frac{y}{a}\right)\right)adx. \tag{2.25}$$

On the other hand, If $a<0$, then $dy = a \times dx$, but in this case the frontiers of the integrals switches as in

$$h(\tilde{Y}) = -\int_{\infty}^{-\infty} \frac{1}{-a}f_{\tilde{X}}\left(\frac{y}{a}\right)\log\left(\frac{1}{-a}f_{\tilde{X}}\left(\frac{y}{a}\right)\right)adx$$

where using the property $\int_a^b(\cdot) = -\int_b^a(\cdot)$, we get

$$h(\tilde{Y}) = -\int_{-\infty}^{\infty} \frac{1}{a} f_{\tilde{X}}\left(\frac{y}{a}\right) \log\left(\frac{1}{-a} f_{\tilde{X}}\left(\frac{y}{a}\right)\right) a dx. \tag{2.26}$$

When (2.25) and (2.26) are considered together, we can write

$$h(\tilde{Y}) = -\int_{-\infty}^{\infty} \frac{1}{a} f_{\tilde{X}}\left(\frac{y}{a}\right) \log\left(\frac{1}{|a|} f_{\tilde{X}}\left(\frac{y}{a}\right)\right) a dx \tag{2.27}$$

where using $y = ax$ and doing the cancellations, we get

$$h(\tilde{Y}) = -\int_{-\infty}^{\infty} f_{\tilde{X}}(x) \log\left(\frac{1}{|a|} f_{\tilde{X}}(x)\right) dx. \tag{2.28}$$

Equation (2.28) can be written as

$$h(\tilde{Y}) = -\int_{-\infty}^{\infty} f_{\tilde{X}}(x)[-\log(|a|) + \log(f_{\tilde{X}}(x))]dx \tag{2.29}$$

which can be simplified as

$$h(\tilde{Y}) = h(\tilde{X}) + \log|a|. \tag{2.30}$$

2.2.1 Properties for Differential Entropy

(1) For the joint differential entropy of N continuous random variables, we have

$$h(\tilde{X}_1, \tilde{X}_2, \ldots, \tilde{X}_N) = \sum_{i=1}^{N} h(\tilde{X}_i|\tilde{X}_{i-1}, \tilde{X}_{i-2}, \ldots, \tilde{X}_1) \tag{2.31}$$

(2) A more general form of $h(\tilde{X}_1, \tilde{X}_2) \leq h(\tilde{X}_1) + h(\tilde{X}_2)$ is

$$h(\tilde{X}_1, \tilde{X}_2, \ldots, \tilde{X}_N) \leq \sum_{i=1}^{N} h(\tilde{X}_i) \tag{2.32}$$

where equality occurs, if the random variables $\tilde{X}_i$ are independent of each other.

(3) We have the inequalities

$$h(\tilde{X}|\tilde{Y}) \leq h(\tilde{X}) \quad h(\tilde{X}|\tilde{Y},\tilde{Z}) \leq h(\tilde{X},\tilde{Z}) \tag{2.33}$$

(4) Differential entropy is invariant to mean shifting, i.e.,

$$h(\tilde{X}+c) = h(\tilde{X}) \tag{2.34}$$

where c is a constant.

(5)

$$h(F(\tilde{X})) \neq h(\tilde{X}) \tag{2.35}$$

where $F(\cdot)$ is a function.

2.2.2 *Conditional Mutual Information for Continuous Random Variables*

The conditional mutual information between $\tilde{X}$ and $\tilde{Y}$ given $\tilde{Z}$ is defined as

$$\begin{aligned} I(\tilde{X};\tilde{Y}|\tilde{Z}) &= E\left\{\log\left(\frac{f(\tilde{X},\tilde{Y}|\tilde{Z})}{f(\tilde{X}|\tilde{Z})f(\tilde{Y}|\tilde{Z})}\right)\right\} \\ &= \int f(x,y,z)\log\frac{f(x,y|z)}{f(x|z)f(y|z)}dxdydz \end{aligned} \tag{2.36}$$

Properties for Conditional Mutual Information

For the conditional mutual information $I(\tilde{X};\tilde{Y}|\tilde{Z})$, we have the following properties.

(1)

$$I(\tilde{X};\tilde{Y}|\tilde{Z}) = h(\tilde{X}|\tilde{Z}) + h(\tilde{Y}|\tilde{Z}) - h(\tilde{X},\tilde{Y}|\tilde{Z}) \tag{2.37}$$

(2)

$$\begin{aligned} I(\tilde{X};\tilde{Y}|\tilde{Z}) &= h(\tilde{X}|\tilde{Z}) - h(\tilde{X}|\tilde{Y},\tilde{Z}) \\ &= h(\tilde{Y}|\tilde{Z}) - h(\tilde{Y}|\tilde{X},\tilde{Z}) \end{aligned} \tag{2.38}$$

(3)

$$I(\tilde{X};\tilde{Y}|\tilde{Z}) \geq 0 \tag{2.39}$$

where equality occurs if $\tilde{X}$ and $\tilde{Y}$ are conditionally independent given $\tilde{Z}$.

(4) $$I(\tilde{X};\tilde{Y},\tilde{Z}) = I(\tilde{X};\tilde{Y}) + I(\tilde{X};\tilde{Z}|\tilde{Y}) \tag{2.40}$$

where $I(\tilde{X};\tilde{Y},\tilde{Z})$ is the mutual information between $\tilde{X}$ and $(\tilde{Y},\tilde{Z})$, on the other hand, $I(\tilde{X};\tilde{Z}|\tilde{Y})$ is the mutual information between $\tilde{X}$ and $\tilde{Z}$ given $\tilde{Y}$.
Note: $\tilde{X}$ and $\tilde{Y}$ are independent if

$$f(x,y) = f(x)f(y). \tag{2.41}$$

On the other hand, $\tilde{X}$ and $\tilde{Y}$ are conditionally independent given $\tilde{Z}$ if

$$f(x,y|z) = f(x|z)f(y|z). \tag{2.42}$$

Example 2.8 Show that

$$I(\tilde{X};\tilde{Y}|\tilde{Z}) = h(\tilde{X}|\tilde{Z}) + h(\tilde{Y}|\tilde{Z}) - h(\tilde{X},\tilde{Y}|\tilde{Z}) \tag{2.43}$$

Solution 2.8 We can start with the definition of conditional mutual information and proceed as follows

$$\begin{aligned} I(\tilde{X};\tilde{Y}|\tilde{Z}) &= E\left\{\log\left(\frac{f(\tilde{X},\tilde{Y}|\tilde{Z})}{f(\tilde{X}|\tilde{Z})f(\tilde{Y}|\tilde{Z})}\right)\right\} \\ &= \underbrace{E\{-\log f(\tilde{X}|\tilde{Z})\}}_{h(\tilde{X}|\tilde{Z})} + \underbrace{E\{-\log f(\tilde{Y}|\tilde{Z})\}}_{h(\tilde{Y}|\tilde{Z})} + \underbrace{E\{\log f(\tilde{X},\tilde{Y}|\tilde{Z})\}}_{h(\tilde{X},\tilde{Y}|\tilde{Z})} \\ &= h(\tilde{X}|\tilde{Z}) + h(\tilde{Y}|\tilde{Z}) - h(\tilde{X},\tilde{Y}|\tilde{Z}). \end{aligned}$$

Thus, we showed that

$$I(\tilde{X};\tilde{Y}|\tilde{Z}) = h(\tilde{X}|\tilde{Z}) + h(\tilde{Y}|\tilde{Z}) - h(\tilde{X},\tilde{Y}|\tilde{Z}).$$

2.2.3 Data Processing Inequality for Continuous Random Variables

If $\tilde{X},\tilde{Y}$ and $\tilde{Z}$ form a Markov chain, i.e., $\tilde{X}$ and $\tilde{Z}$ are conditionally independent given $\tilde{Y}$, then

$$I(\tilde{X};\tilde{Y}) \geq I(\tilde{X};\tilde{Z}). \tag{2.44}$$

Proof Due to the conditional independence of $\tilde{X}$ and $\tilde{Z}$ given $\tilde{Y}$, we have $I(\tilde{X};\tilde{Z}|\tilde{Y}) = 0$. We can write the mutual information $I(\tilde{X};\tilde{Y},\tilde{Z})$ as

$$I(\tilde{X};\tilde{Y},\tilde{Z}) = I(\tilde{X};\tilde{Y}) + \underbrace{I(\tilde{X};\tilde{Z}|\tilde{Y})}_{=0} \rightarrow I(\tilde{X};\tilde{Y},\tilde{Z}) = I(\tilde{X};\tilde{Y}). \tag{2.45}$$

Alternatively, $I(\tilde{X};\tilde{Y},\tilde{Z})$ can also be written as

$$I(\tilde{X};\tilde{Y},\tilde{Z}) = I(\tilde{X};\tilde{Z}) + I(\tilde{X};\tilde{Y}|\tilde{Z}). \tag{2.46}$$

Equating the right hand sides of (2.45) and (2.46), we obtain

$$I(\tilde{X};\tilde{Y}) = I(\tilde{X};\tilde{Z}) + I(\tilde{X};\tilde{Y}|\tilde{Z}). \tag{2.47}$$

Since $I(\cdot) \geq 0$, from (2.47), we can write

$$I(\tilde{X};\tilde{Y}) \geq I(\tilde{X};\tilde{Z}). \tag{2.48}$$

Example 2.9 Let $\tilde{X}$ be a continuous random variable, and $\tilde{N} \sim N(0, \sigma_N^2)$, $\tilde{Y} = \tilde{X} + \tilde{N}$. The channel whose input and output are $\tilde{X}$ and $\tilde{Y}$ is called the additive white Gaussian noise (AWGN) channel which is depicted in Fig. 2.1. If $\tilde{X}$ is chosen as $\tilde{X} \sim N(0, \sigma_X^2)$ and assuming also that $\tilde{X}$ and $\tilde{N}$ are independent of each other, then it can be shown that the mutual information between $\tilde{X}$ and $\tilde{Y}$ can be calculated as

$$I(\tilde{X};\tilde{Y}) = \frac{1}{2}\log\left(1 + \frac{\sigma_X^2}{\sigma_N^2}\right). \tag{2.49}$$

Using (2.49), show that for the system given in Fig. 2.2 where $\tilde{N}_1 \sim N(0, \sigma_N^2)$, $\tilde{N}_2 \sim N(0, \sigma_N^2)$, $\tilde{N}_1$ and $\tilde{N}_2$ are independent random variables, we have

$$I(\tilde{X};\tilde{Z}) \leq I(\tilde{X};\tilde{Y}).$$

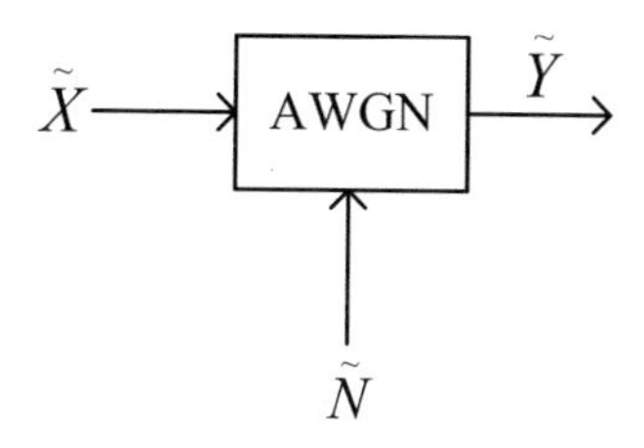

Fig. 2.1 AWGN channel

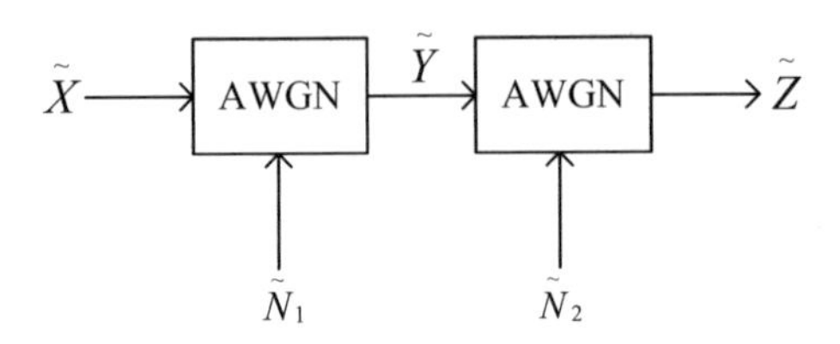

Fig. 2.2 Cascaded AWGN channels

Solution 2.9 Considering Fig. 2.2, we can write

$$\tilde{Y} = \tilde{X} + \tilde{N}_1 \quad \text{and} \quad \tilde{Z} = \tilde{Y} + \tilde{N}_2 \to \tilde{Z} = \tilde{X} + \underbrace{\tilde{N}_1 + \tilde{N}_2}_{\tilde{N}_3}. \tag{2.50}$$

Since $\tilde{N}_1$ and $\tilde{N}_2$ are independent random variables, considering (2.50), we can write that

$$\sigma^2_{N_3} = \sigma^2_{N_1} + \sigma^2_{N_2} \to \sigma^2_{N_3} = \sigma^2_N + \sigma^2_N \to \sigma^2_{N_3} = 2\sigma^2_N.$$

The mutual information between $\tilde{X}$ and $\tilde{Y}$ can be calculated as

$$I(\tilde{X};\tilde{Y}) = \frac{1}{2}\log\left(1 + \frac{\sigma^2_X}{\sigma^2_{N_1}}\right) \to I(\tilde{X};\tilde{Y}) = \frac{1}{2}\log\left(1 + \frac{\sigma^2_X}{\sigma^2_N}\right). \tag{2.51}$$

The mutual information between $\tilde{X}$ and $\tilde{Z}$ is calculated as

$$I(\tilde{X};\tilde{Z}) = \frac{1}{2}\log\left(1 + \frac{\sigma^2_X}{\sigma^2_{N_3}}\right) \to I(\tilde{X};\tilde{Z}) = \frac{1}{2}\log\left(1 + \frac{\sigma^2_X}{2\sigma^2_N}\right). \tag{2.52}$$

If (2.51) and (2.52) are compared to each other, we see that

$$I(\tilde{X};\tilde{Z}) = \frac{1}{2}\log\left(1 + \frac{\sigma^2_X}{2\sigma^2_N}\right) < I(\tilde{X};\tilde{Y}) = \frac{1}{2}\log\left(1 + \frac{\sigma^2_X}{\sigma^2_N}\right). \tag{2.53}$$

Exercise For the communication system in Fig. 2.3, we have $\tilde{X} \sim N(0, \sigma^2_X)$, $\tilde{N}_1 \sim N(0, \sigma^2_N)$, $\tilde{N}_2 \sim N(0, \sigma^2_N)$ and $\tilde{X}, \tilde{N}_1, \tilde{N}_2$, and $\tilde{N}_3$ are independent of each other. Show that

$$I(\tilde{X};\tilde{Z}) \le I(\tilde{X};\tilde{W}) \le I(\tilde{X};\tilde{Y})$$

Exercise For the previous exercise, consider what happens as the number of AWGN channels goes to infinity.

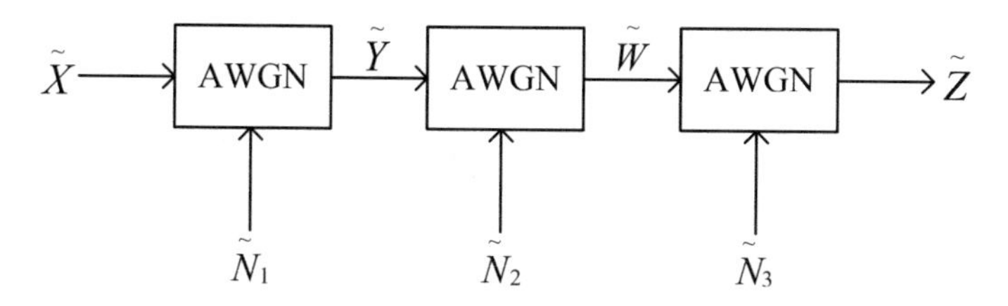

Fig. 2.3 Cascaded AWGN channels

Example 2.10 $\tilde{X}$ is a continuous random variable with variance σ^2, and $\tilde{Y}$ is a Gaussian distributed random variable with the same variance, i.e., $\tilde{Y} \sim N(\mu, \sigma^2)$, and $f(x)$ and $g(y)$ are the probability density functions of $\tilde{X}$ and $\tilde{Y}$ respectively. Show that,

$$\int g(y) \log g(y) dy = \int f(x) \log g(x) dx.$$

Solution 2.10 For the given distributions we have

$$\int g(y) dy = \int f(x) dx = 1.$$

The Gaussian random variable has the distribution

$$g(y) = \frac{1}{\sqrt{2\pi\sigma^2}} e^{-\frac{(y-\mu)^2}{2\sigma^2}}. \tag{2.54}$$

When (2.54) is used inside the logarithmic expression in

$$\int g(y) \log g(y) dy$$

we obtain

$$\int g(y) \log\left(\frac{1}{\sqrt{2\pi\sigma^2}} e^{-\frac{(y-\mu_y)^2}{2\sigma^2}}\right) dy$$

which can be written as

$$\int g(y) \left[\log\left(\frac{1}{\sqrt{2\pi\sigma^2}}\right) - \frac{1}{\ln 2} \frac{(y-\mu_y)^2}{2\sigma^2}\right] dy. \tag{2.55}$$

Equation (2.55) is expanded as

$$\log\left(\frac{1}{\sqrt{2\pi\sigma^2}}\right)\underbrace{\int g(y)dy}_{=\int f(x)dx=1}-\frac{1}{2\sigma^2\ln 2}\underbrace{\int (y-\mu_y)^2 g(y)dy}_{\int (x-\mu_x)^2 f(x)dx}$$

where substituting

$$\int f(x)dx$$

for

$$\int g(y)dy$$

and substituting

$$\int (x-\mu_x)^2 f(x)dx$$

for

$$\int (y-\mu_y)^2 g(y)dy$$

we obtain

$$\log\left(\frac{1}{\sqrt{2\pi\sigma^2}}\right)\int f(x)dx-\frac{1}{2\sigma^2\ln 2}\int (x-\mu_x)^2 f(x)dx \tag{2.56}$$

which can be written as

$$\int f(x)\underbrace{\left(\log\left(\frac{1}{\sqrt{2\pi\sigma^2}}\right)-\frac{(x-\mu_x)^2}{2\sigma^2\ln 2}\right)}_{\log g(x)}dx. \tag{2.57}$$

Equation (2.57) can be written in more compact form as

$$\int f(x)\log g(x)dx. \tag{2.58}$$

Thus, we showed that

$$\int g(y)\log g(y)dy = \int f(x)\log g(x)dx.$$

Example 2.11 Show that the entropy of any continuous random variable $\tilde{X}$ with variance σ^2 is smaller than or equal to the entropy of the random variable with Gaussian distribution having variance σ^2, i.e.,

$$h(\tilde{X}) \le h_g \to h(\tilde{X}) \le \frac{1}{2}\log(2\pi\sigma^2)$$

where equality occurs if $\tilde{X}$ is also Gaussian distributed.

Solution 2.11 Let $\tilde{Y} \sim N(\mu_y, \sigma^2)$, i.e., $\tilde{Y}$ is a zero mean Gaussian distributed continuous random variable. Then, the entropy of $\tilde{Y}$ happens to be

$$h(\tilde{Y}) = -\int p(y)\log p(y)dy \to h(\tilde{Y}) = \frac{1}{2}\log(2\pi\sigma^2)$$

where

$$p(y) = \frac{1}{\sqrt{2\pi\sigma^2}}\mathrm{e}^{-\frac{(y-\mu_y)^2}{2\sigma^2}}. \tag{2.59}$$

Let's consider the difference

$$h(\tilde{Y}) - h(\tilde{X}) = -\int p(y)\log p(y)dy + \int f(x)\log f(x)dx \tag{2.60}$$

where employing the equality

$$\int g(y)\log g(y)dy = \int f(x)\log g(x)dx \tag{2.61}$$

obtained in our previous example, we get

$$h(\tilde{Y}) - h(\tilde{X}) = -\int f(x)\log g(x)dx + \int f(x)\log f(x)dx \tag{2.62}$$

which can be written as

$$h(\tilde{Y}) - h(\tilde{X}) = \int f(x) \log \frac{f(x)}{g(x)} dx. \tag{2.63}$$

Equation (2.63) can be expressed as

$$h(\tilde{Y}) - h(\tilde{X}) = D[f(x)||g(x)]. \tag{2.64}$$

Since $D[\cdot] \geq 0$, we have

$$h(\tilde{Y}) - h(\tilde{X}) \geq 0$$

which means

$$h(\tilde{X}) \leq h(\tilde{Y}) = \frac{1}{2}\log(2\pi\sigma^2) \rightarrow h(\tilde{X}) \leq \frac{1}{2}\log(2\pi\sigma^2). \tag{2.65}$$

Exercise Let the continuous random variable $\tilde{X}$ have exponential distribution

$$f(x) = \begin{cases} \lambda e^{-\lambda x} & x \geq 0 \\ 0 & x < 0. \end{cases}$$

The variance of the exponential random variable is

$$Var(\tilde{X}) = \frac{1}{\lambda^2}.$$

Calculate the entropy of $\tilde{X}$ for $\lambda = 2$. Let $\tilde{Y} \sim N(0, 1/\lambda^2)$, calculate the entropy of $\tilde{Y}$ for $\lambda = 2$, and compare it to the calculated entropy of $\tilde{X}$.

Exercise $\tilde{X}$ is a Gaussian random variable with variance $\sigma^2 = 4$ and $\tilde{Y} = 2\tilde{X}$. Calculate the entropies of $\tilde{X}$ and $\tilde{Y}$.

Example 2.12 Show that for continuous random variables defined on a finite interval $[a\ b]$, the uniformly distributed random variable has maximum entropy.

Solution 2.12 Let the continuous random variable $\tilde{X}_1$ be uniformly distributed on $[ab]$, and its probability density function be denoted by $f(x)$. It is clear that

$$f(x) = \frac{1}{b-a} \quad a \leq x \leq b.$$

The entropy of $\tilde{X}_1$ can be found as

$$H(\tilde{X}_1) = \log(b - a).$$

Let the continuous random variable $\tilde{X}_2$ has the distribution $g(x)$, we want to show that

$$H(\tilde{X}_2) \leq H(\tilde{X}_1).$$

Consider the difference

$$H(\tilde{X}_2) - H(\tilde{X}_1)$$

which can be explicitly written as

$$-\int g(x) \log g(x) dx - \log(b - a)$$

where substituting

$$\log(b - a) \int g(x) dx$$

for

$$\log(b - a)$$

we obtain

$$-\int g(x) \log g(x) dx - \int \log(b - a) g(x) dx$$

in which using

$$f(x)$$

for

$$\frac{1}{b - a}$$

we get

$$-\int g(x) \log g(x) dx + \int g(x) \log f(x) dx$$

leading to

$$-\int g(x)\log\frac{g(x)}{f(x)}dx$$

which can be identified as

$$-D(g(x)||f(x))$$

where $D(g(x)||f(x))$ is a non-negative quantity, i.e., $-D(g(x)||f(x))\leq 0$ which implies that

$$H(\tilde{X}_2)-H(\tilde{X}_1)\leq 0\rightarrow H(\tilde{X}_2)\leq H(\tilde{X}_1).$$

Example 2.13 Let the continuous random variables $\tilde{X}_1$ and $\tilde{X}_2$ have the probability distribution functions $f(x)$ and $g(x)$. Show that

$$H(\tilde{X}_1)\leq -\int f(x)\log g(x)dx.$$

Solution 2.13 Let's consider the relative entropy between $\tilde{X}_1$ and $\tilde{X}_2$, i.e. $D[f(x)||g(x)]$ defined as

$$D[f(x)||g(x)]=\int f(x)\log\frac{f(x)}{g(x)}dx$$

which can be written as

$$D[f(x)||g(x)]=\int f(x)\log f(x)dx-\int f(x)\log g(x)dx$$

where employing

$$D[f(x)||g(x)]\geq 0$$

we get

$$\int f(x)\log f(x)dx-\int f(x)\log g(x)dx\geq 0\rightarrow -H(\tilde{X})-\int f(x)\log g(x)dx\geq 0$$

leading to

$$H(\tilde{X}) \leq -\int f(x) \log g(x) dx.$$

Example 2.14 Show that for continuous random variables having the same mean value, the exponentially distributed random variable has maximum entropy.

Solution 2.14 Let the continuous random variable $\tilde{X}_1$ have exponential distribution, i.e., its probability density function can be written as

$$f(x) = \lambda e^{-\lambda x}.$$

It can be shown that

$$E(\tilde{X}_1) = \frac{1}{\lambda} \quad H(\tilde{X}_1) = \log \frac{1}{\lambda} + 1.$$

The continuous random variable $\tilde{X}_2$ has the probability density function $g(x)$, and the mean value of $\tilde{X}_2$ is

$$E(\tilde{X}_2) = \frac{1}{\lambda}$$

i.e., it has the same mean value as $\tilde{X}_1$. We want to show that

$$H(\tilde{X}_2) \leq H(\tilde{X}_1).$$

From Example 2.13, we have

$$H(\tilde{X}_2) \leq -\int g(x) \log f(x) dx$$

where employing $f(x) = \lambda e^{-\lambda x}$, we obtain

$$H(\tilde{X}_2) \leq -\int g(x) \log \lambda e^{-\lambda x} dx$$

which can be manipulated as

$$\begin{aligned}
H(\tilde{X}_2) &\leq -\int g(x)\log \lambda e^{-\lambda x} dx \\
&\leq -\int g(x)\left(\log \lambda + \log e^{-\lambda x}\right) dx \\
&\leq \int g(x)\left(\log \frac{1}{\lambda} - \log e^{-\lambda x}\right) dx \\
&\leq \log \frac{1}{\lambda} + \lambda \underbrace{\int x g(x) dx}_{1/\lambda} \\
&\leq \underbrace{\log \frac{1}{\lambda} + 1}_{H(\tilde{X}_1)}
\end{aligned}$$

leading to

$$H(\tilde{X}_2) \leq H(\tilde{X}_1).$$

Theorem 2.5 *Let $\tilde{X}$ be a continuous random variable, and $\tilde{X}_a$ be the estimation of $\tilde{X}$. The estimation error $E(\tilde{X} - \tilde{X}_a)^2$ satisfy the bound*

$$E(\tilde{X} - \tilde{X}_a)^2 \geq \frac{1}{2\pi e} 2^{2h(\tilde{X})}.$$

Proof 2.5 It is known that for equal variance random variables, the Gaussian random variable has the largest entropy, and the entropy $h(\tilde{X})$ of any random variable with variance σ^2 is smaller than the entropy of the Gaussian random variable, i.e.,

$$h(\tilde{X}) \leq \frac{1}{2}\log(2\pi e \sigma^2)$$

from which we obtain

$$\frac{1}{2\pi e} 2^{2h(\tilde{X})} \leq \sigma^2.$$

The estimation error

$$E(\tilde{X} - \tilde{X}_a)^2$$

satisfy

$$E(\tilde{X} - \tilde{X}_a)^2 \geq E(\tilde{X} - E(\tilde{X}))^2 = \sigma^2$$

in which employing $\frac{1}{2\pi e} 2^{2h(\tilde{X})} \leq \sigma^2$, we get

$$E(\tilde{X} - \tilde{X}_a)^2 \geq \frac{1}{2\pi e} 2^{2h(\tilde{X})}.$$

2.3 Channel Capacity

Before explaining the channel capacity, let's give the definition of discrete memoryless channel.

Definition Discrete memoryless channel:
A discrete channel has an input symbol set $R_{\tilde{X}}$ and an output symbol set $R_{\tilde{Y}}$. The symbols at the input of the discrete channel are generated by a discrete random variable $\tilde{X}$, and similarly the symbols at the output of the discrete memoryless channel are generated by a discrete random variable $\tilde{Y}$. The discrete random variables $\tilde{X}$ and $\tilde{Y}$ have joint distribution $p(x, y)$, and the conditional probability $p(y|x)$ is called the channel transition probabilities.

In short, a discrete memoryless channel is nothing but the conditional probability function $p(y|x)$ defined between two discrete random variables $\tilde{X}$ and $\tilde{Y}$.

Definition Channel capacity:
The capacity of a communication channel is defined as

$$C = \max_{p(x)} I(\tilde{X}; \tilde{Y}) \tag{2.66}$$

where the maximization is performed considering all possible input distributions. Channel capacity can be calculated for both discrete and continuous communication channels, i.e., it can be calculated for both discrete and continuous random variables.

In Fig. 2.4 graphical illustration of the channel capacity is shown. However, the graph in Fig. 2.4 is not a complete graph. It is a two dimensional graph, however, depending on the number of symbols in input distribution, the horizontal axis may have many dimensions.

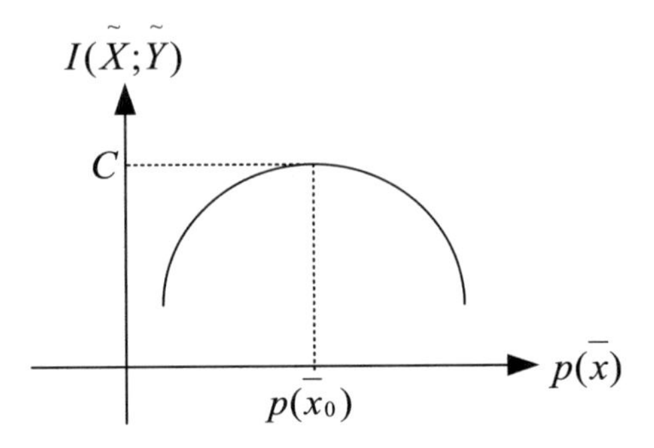

Fig. 2.4 Channel capacity illustration

2.3.1 Discrete Channel Capacity

We will first study discrete memoryless channel capacity, then focus on the continuous channel capacity. As we mentioned in the definition, channel capacity is calculated using

$$C = \max_{p(x)} I(\tilde{X};\tilde{Y})$$

where $I(\tilde{X};\tilde{Y})$ can be calculated using either

$$I(\tilde{X};\tilde{Y}) = H(\tilde{X}) - H(\tilde{X}|\tilde{Y}) \tag{2.67}$$

or

$$I(\tilde{X};\tilde{Y}) = H(\tilde{Y}) - H(\tilde{Y}|\tilde{X}) \tag{2.68}$$

depending on the structure of the channel. For some channels (2.67) may be more useful for capacity calculation, whereas for others (2.68) can be the preferred choice.

Then, from (2.67) and (2.68), which one should be chosen for the calculation of channel capacity? The answer of this question is as follows. We should carefully inspect the channel very well, and while deciding on (2.67) or (2.68), we should try to see the one that uses less calculation and is easier to manipulate mathematically.

Properties of Discrete Channel Capacity

Let's state some properties of the discrete channel capacity.

(1) $C \geq 0$, since $C = \max_{p(x)} I(\tilde{X};\tilde{Y})$ and $I(\tilde{X};\tilde{Y}) \geq 0$
(2) $C \leq \log|R_{\tilde{X}}|$ and $C \leq \log|R_{\tilde{Y}}|$ where $|R_{\tilde{X}}|$ and $|R_{\tilde{Y}}|$ are the number of elements in the range sets of $\tilde{X}$ and $\tilde{Y}$.

Proof Employing (2.67) in $C = \max_{p(x)} I(\tilde{X}; \tilde{Y})$ we obtain

$$C = \max_{p(x)} \left[H(\tilde{X}) - H(\tilde{X}|\tilde{Y}) \right] \tag{2.69}$$

where considering $H(\tilde{X}|\tilde{Y}) \geq 0$ we can write

$$C \leq \max_{p(x)} \left[H(\tilde{X}) \right] \tag{2.70}$$

in which employing

$$H(\tilde{X}) \leq \log |R_{\tilde{X}}|$$

we get

$$C \leq \log \left| R_{\tilde{X}} \right|.$$

In a similar manner, if we proceed with (2.68) in $C = \max_{p(x)} I(\tilde{X}; \tilde{Y})$, we get

$$C \leq \log \left| R_{\tilde{Y}} \right|.$$

We mentioned previously that $I(\tilde{X}; \tilde{Y})$ is a concave function of $p(x)$, i.e., probability mass function of $\tilde{X}$. Then, finding the capacity expressed as

$$C = \max_{p(x)} I(\tilde{X}; \tilde{Y})$$

is nothing but finding the local global maximum of the mutual information function $I(\tilde{X}; \tilde{Y})$. This is nothing but an optimization problem. Optimum value of some functions can be found mathematically, and for some others, search algorithms are run to find the closest value to the optimal value, and for the rest, just computer trials are performed, no direct or indirect mathematical analysis may be possible.In addition, $I(\tilde{X}; \tilde{Y})$ is a continuous function of $p(x)$. This means that for close set of probability values, we do not see a sharp change in $I(\tilde{X}; \tilde{Y})$. There is either a smooth increment or decrement observed in the graph of $I(\tilde{X}; \tilde{Y})$.Let's now solve some problems illustrating the capacity calculation for discrete memoryless channels.

Example 2.15 For the binary symmetric channel shown in Fig. 2.5, is it possible for the output to have uniform distribution? If so, for which input distributions, we get uniform distribution at the output.

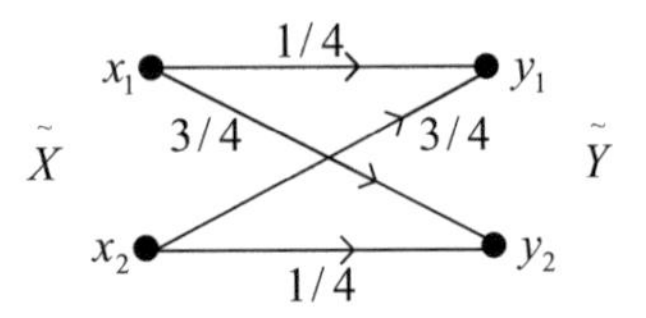

Fig. 2.5 Binary symmetric channel for Example 2.15

Solution 2.15 Let the input distribution be as in

$$p(x_1) = \alpha \quad p(x_2) = 1 - \alpha.$$

Then, considering the transition probabilities $p(y|x)$ given in the Fig. 2.5, we can calculate the output distribution using

$$p(y) = \sum_x p(x, y) \rightarrow p(y) = \sum_x p(y|x)p(x)$$

as in

$$\begin{aligned} p(y_1) &= p(y_1|x_1)p(x_1) + p(y_1|x_2)p(x_2) \rightarrow p(y_1) = \frac{\alpha}{4} + \frac{3(1-\alpha)}{4} \\ p(y_2) &= p(y_2|x_1)p(x_1) + p(y_2|x_2)p(x_2) \rightarrow p(y_2) = \frac{3\alpha}{4} + \frac{(1-\alpha)}{4} \end{aligned} \tag{2.71}$$

For the output random variable to have uniform distribution, we should have

$$p(y_1) = p(y_2) = \frac{1}{2} \tag{2.72}$$

Equating the first equation of (2.71) to 1/2, we get

$$\frac{\alpha}{4} + \frac{3(1-\alpha)}{4} = \frac{1}{2}$$

whose solution is

$$\alpha = \frac{1}{2}.$$

If $\alpha = 1/2$ is substituted into second equation of (2.71), we get also $p(y_2) = 1/2$. Thus, $p(y_1) = p(y_2) = 1/2$ for $\alpha = 1/2$. This means that, uniform input distribution produces uniform input distribution at the output of the binary symmetric channel.

We can conclude that it is possible to have a uniformly distributed random variable at the output of a discrete binary symmetric channel.

Exercise For the binary symmetric channel shown in Fig. 2.6, show that it is possible to have uniformly distributed random variable at the output of the channel for some input random variables. Determine the distribution of the input random variable that yields uniformly distributed random variable at the output of the binary symmetric channel.

Exercise For the binary communication channel shown in Fig. 2.7, is it possible for the output random variable to have uniform distribution? If yes, for which input distributions, we get uniform distribution at the output.

Exercise For the binary channel with unequal transition probabilities shown in Fig. 2.8, is it possible to have uniformly distributed random variable at the output of the channel for some input random variables? If yes, under which criteria, it is possible.

Example 2.16 Calculate the channel capacity of the binary symmetric channel shown in Fig. 2.9.

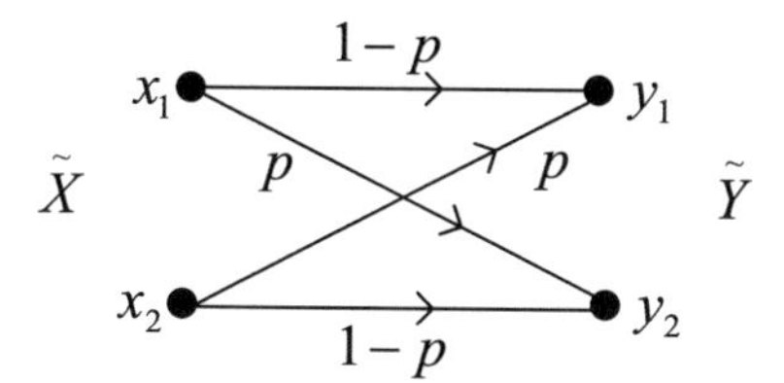

Fig. 2.6 Binary symmetric channel exercise

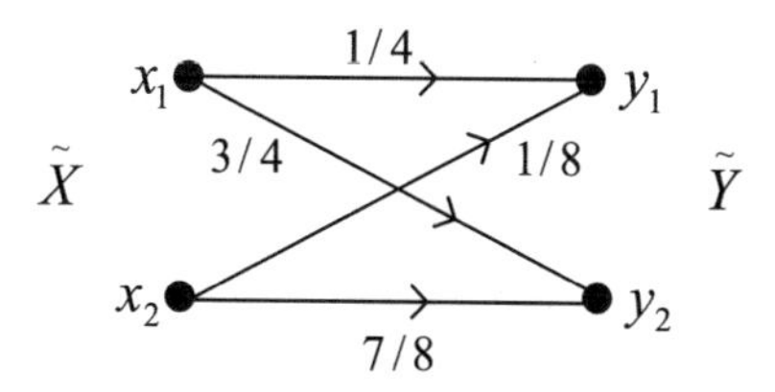

Fig. 2.7 Binary communication channel for exercise

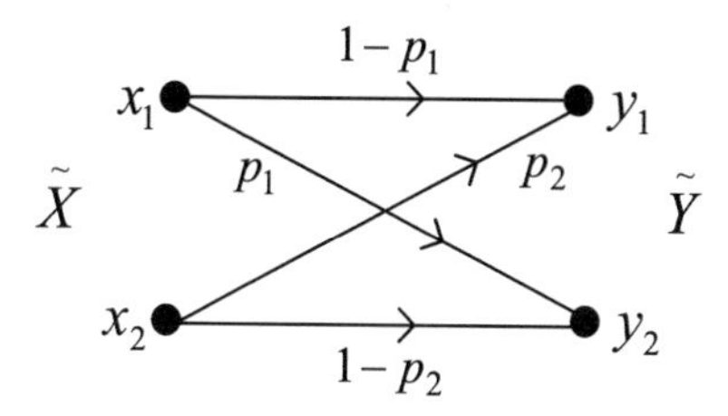

Fig. 2.8 Binary communication channel for exercise

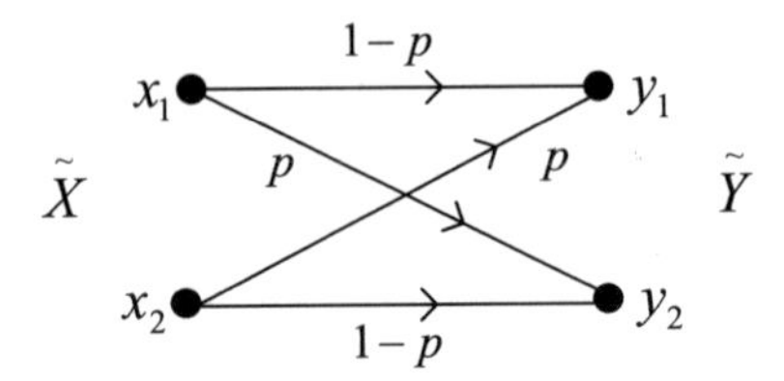

Fig. 2.9 Binary symmetric channel for Example 2.16

Solution 2.16 To calculate the channel capacity

$$C = \max_{p(x)} I(\tilde{X};\tilde{Y})$$

we should first calculate the mutual information $I(\tilde{X};\tilde{Y})$ between channel input and output, and next try to optimize it if necessary. The mutual information $I(\tilde{X};\tilde{Y})$ can be calculated using either

$$I(\tilde{X};\tilde{Y}) = H(\tilde{Y}) - H(\tilde{Y}|\tilde{X})$$

or

$$I(\tilde{X};\tilde{Y}) = H(\tilde{X}) - H(\tilde{X}|\tilde{Y}).$$

For this example, $I(\tilde{X};\tilde{Y}) = H(\tilde{Y}) - H(\tilde{Y}|\tilde{X})$ is more suitable. However, the other expression $I(\tilde{X};\tilde{Y}) = H(\tilde{X}) - H(\tilde{X}|\tilde{Y})$ can also be utilized.
Let's fist use

$$I(\tilde{X};\tilde{Y}) = H(\tilde{Y}) - H(\tilde{Y}|\tilde{X})$$

to calculate the mutual information. In

$$I(\tilde{X};\tilde{Y}) = H(\tilde{Y}) - H(\tilde{Y}|\tilde{X})$$

the conditional entropy $H(\tilde{Y}|\tilde{X})$ can be calculated using

$$H(\tilde{Y}|\tilde{X}) = \sum_{x} p(x)H(\tilde{Y}|x)$$

where

$$H(\tilde{Y}|x) = -\sum_{y} p(y|x)\log p(y|x)$$

which can be calculated for $x = x_1$ as

$$H(\tilde{Y}|x = x_1) = -\sum_y p(y|x = x_1)\log p(y|x = x_1) \rightarrow$$

$$H(\tilde{Y}|x = x_1) = -[p(y = y_1|x = x_1)\log p(y = y_1|x = x_1) + p(y = y_2|x = x_1)\log p(y = y_2|x = x_1)] \rightarrow$$

$$H(\tilde{Y}|x = x_1) = -[(1 - p)\log(1 - p) + p\log(p)]$$

which is a function of p only, for his reason, we can denote $H(\tilde{Y}|x = x_1)$ by the special notation $H_b(p)$, i.e., $H(\tilde{Y}|x = x_1) = H_b(p)$.
The calculation of

$$H(\tilde{Y}|x) = -\sum_y p(y|x)\log p(y|x)$$

for $x = x_2$ gives the same result, i.e., $H(\tilde{Y}|x = x_2) = H_b(p)$. Now, we can utilize

$$H(\tilde{Y}|\tilde{X}) = \sum_x p(x)H(\tilde{Y}|x) \tag{2.73}$$

to calculate the conditional entropy. Expanding (2.73) for all x values, we get

$$H(\tilde{Y}|\tilde{X}) = p(x_1)H(\tilde{Y}|x_1) + p(x_2)H(\tilde{Y}|x_2)$$

where substituting $H(\tilde{Y}|x = x_1) = H_b(p)$ and $H(\tilde{Y}|x = x_2) = H_b(p)$, we obtain

$$H(\tilde{Y}|\tilde{X}) = p(x_1)H_b(p) + p(x_2)H_b(p) \rightarrow H(\tilde{Y}|\tilde{X}) = \Bigg[\underbrace{p(x_1) + p(x_2)}_{=1}\Bigg]H_b(p).$$

Hence, we get

$$H(\tilde{Y}|\tilde{X}) = H_b(p). \tag{2.74}$$

Using the obtained result (2.74) in

$$I(\tilde{X};\tilde{Y}) = H(\tilde{Y}) - H(\tilde{Y}|\tilde{X}),$$

we get

$$I(\tilde{X};\tilde{Y}) = H(\tilde{Y}) - H_b(p).$$

Now, we will try to find capacity, i.e., try to find

$$C = \max_{p(x)} I(\tilde{X}; \tilde{Y}) \rightarrow C = \max_{p(x)} \left[H(\tilde{Y}) - H_b(p) \right]. \tag{2.75}$$

Since $H_b(p)$ is a constant value, (2.75) can be written as

$$C = \max_{p(x)} \left[H(\tilde{Y}) \right] - H_b(p). \tag{2.76}$$

A discrete random variable gets its maximum entropy if it has uniform distribution. In our previous examples, we showed that it is possible for the random variable at the output of a binary symmetric channel to have uniform distribution, and this is possible if the discrete random variable at the input of the binary symmetric channel has uniform distribution.

If $p(x_1) = 1/2, p(x_2) = 1/2$, then output distribution becomes as $p(y_1) = 1/2$, $p(y_2) = 1/2$, and in this case we get

$$\max\left[H(\tilde{Y})\right] = \log |R_{\tilde{Y}}| \rightarrow \max\left[H(\tilde{Y})\right] = \log 2 \rightarrow \max\left[H(\tilde{Y})\right] = 1.$$

Then, (2.76) happens to be

$$C = 1 - H_b(p)$$

The graphs of $H_b(p)$ and $C = 1 - H_b(p)$ are depicted in Fig. 2.10 where it is seen that capacity becomes equal to 0 when $p = 0.5$ at which $H(\tilde{Y}|\tilde{X}) = H_b(p)$ is maximum, i.e., if uncertainty is maximum, then capacity get its minimum value.

Solution 2 Let's use $I(\tilde{X}; \tilde{Y}) = H(\tilde{X}) - H(\tilde{X}|\tilde{Y})$ for the calculation of mutual information. Let the input distribution be as in

$$p(x_1) = \alpha \quad p(x_2) = 1 - \alpha.$$

Entropy of the $\tilde{X}$ can be calculated as

$$H(\tilde{X}) = -\sum_{x} p(x) \log p(x) \rightarrow H(\tilde{X}) = -[\alpha \log \alpha + (1 - \alpha) \log(1 - \alpha)] \tag{2.77}$$

The entropy expression in (2.77) is a function of α and can be denoted as $H_b(\alpha)$ where b means binary. The conditional entropy $H(\tilde{X}|\tilde{Y})$ can be calculated using

$$H(\tilde{X}|\tilde{Y}) = -\sum_{x,y} p(x, y) \log \frac{p(x, y)}{p(y)}$$

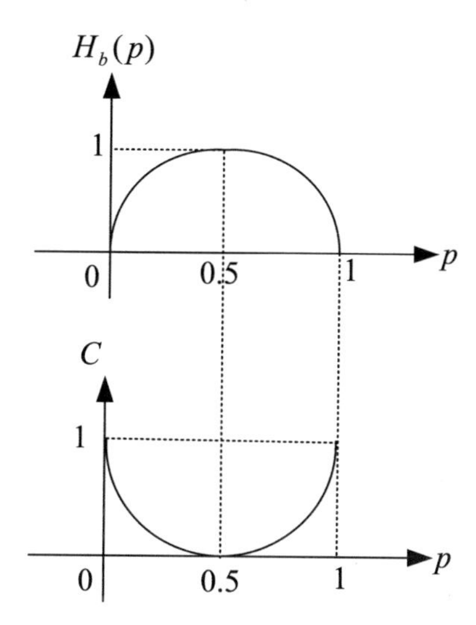

Fig. 2.10 Capacity w.r.t $H_b(p)$

where $p(y)$ and $p(x,y)$ can be calculated using

$$p(y) = \sum_x p(x,y) \rightarrow p(y) = \sum_x p(y|x)p(x) \quad p(x,y) = p(y|x)p(x).$$

leading to

$$\begin{aligned}
p(y_1) &= \alpha(1-p) + (1-\alpha)p \rightarrow p(y_1) = \alpha + p - 2\alpha p \\
p(y_2) &= \alpha p + (1-\alpha)(1-p) \rightarrow p(y_2) = 1 - \alpha - p + 2\alpha p \\
p(x_1,y_1) &= \alpha(1-p) \\
p(x_1,y_2) &= \alpha p \\
p(x_2,y_1) &= (1-\alpha p) \\
p(x_2,y_2) &= (1-\alpha)(1-p).
\end{aligned}$$

Using the found values, we can calculate

$$H(\tilde{X}|\tilde{Y}) = -\sum_{x,y} p(x,y) \log \frac{p(x,y)}{p(y)}$$

as

$$\begin{aligned}
H(\tilde{X}|\tilde{Y}) = -\Bigg[&\alpha(1-p) \log \frac{\alpha(1-p)}{\alpha + p - 2\alpha p} + \alpha p \log \frac{\alpha p}{1 - \alpha - p + 2\alpha p} + (1-\alpha p) \log \frac{(1-\alpha p)}{\alpha + p - 2\alpha p} \\
&+ (1-\alpha)(1-p) \log \frac{(1-\alpha)(1-p)}{1 - \alpha - p + 2\alpha p} \Bigg]
\end{aligned}$$

which is not an easy expression to simplify. Thus, for binary symmetric channel, to get a general expression for the channel capacity, it is better to utilize

$$I(\tilde{X};\tilde{Y}) = H(\tilde{Y}) - H(\tilde{Y}|\tilde{X})$$

rather than

$$I(\tilde{X};\tilde{Y}) = H(\tilde{X}) - H(\tilde{X}|\tilde{Y}).$$

Example 2.17 Calculate the channel capacity of the binary symmetric channel shown in Fig. 2.11 and calculate the value of mutual information between channel input and output when input distribution is $p(x=0) = 1/4, p(x=1) = 3/4$ and compare the calculated mutual information to the channel capacity computed.

Solution 2.17 The channel capacity of the binary symmetric channel shown in Fig. 2.11 can be calculated as

$$C = 1 - H_b\left(p = \frac{1}{4}\right) \rightarrow C = 0.1887.$$

The mutual information between $\tilde{X}$ and $\tilde{Y}$ for the given distribution can be calculated using

$$I(\tilde{X};\tilde{Y}) = H(\tilde{Y}) - H_b\left(p = \frac{1}{4}\right)$$

resulting in

$$I(\tilde{X};\tilde{Y}) = 0.1432.$$

We see that mutual information for the given input distribution is smaller than the capacity of the channel, i.e.,

$$I(\tilde{X};\tilde{Y}) = 0.1432 < C = 0.1887.$$

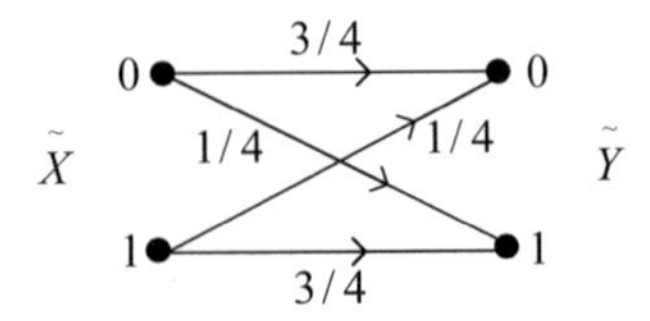

Fig. 2.11 Binary symmetric channel for Example 2.17

Exercise Calculate the channel capacity of the binary symmetric channel shown in Fig. 2.12, and calculate the value of mutual information between channel input and output when input distribution is $p(x=0)=3/8, p(x=1)=5/8$, and compare the calculated mutual information to the channel capacity.

Example 2.18 Calculate the channel capacity of the binary symmetric channel shown in Fig. 2.13, and comment on the meaning of the channel capacity, give some examples to illustrate the meaning of channel capacity.

Solution 2.18 We can calculate the channel capacity using

$$C = 1 - H_b(p) \rightarrow C = 1 + [(1-p)\log(1-p) + p\log(p)]$$

as in

$$C = 1 + \left[\frac{3}{4}\log\left(\frac{3}{4}\right) + \frac{1}{4}\log\left(\frac{1}{4}\right)\right] \rightarrow C = 1 - 0.8113 \rightarrow C = 0.1887 \text{ bits/sym.}$$

Don't forget than entropy, mutual information and, capacity is nothing but probabilistic average values. And according to the law of large numbers, probabilistic average approaches to arithmetic average as the number of trials goes to infinity.

In our case, symbols are also chosen from bits. We found the capacity as $C = 0.1887$ bits/sym. This means that if the bits at the input of the binary symmetric channel is generated according to uniform distribution, i.e., bits '0' and '1' are generated with equal probability 1/2, and transmitted through binary symmetric channel, for every transmitted bit, at most 0.1887 bits will be received correctly in average, the others will be received with error.

What we have explained may not be meaningful in practical applications. Since, we do not have fractional bit in practical applications. However, we can talk about the fractional bits in mathematics, or in theory, there is no limitation for this.

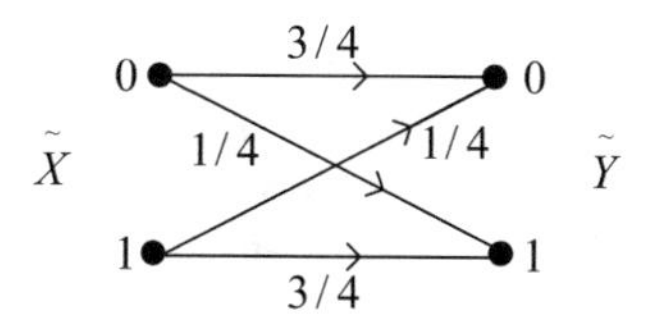

Fig. 2.12 Binary symmetric channel for exercise

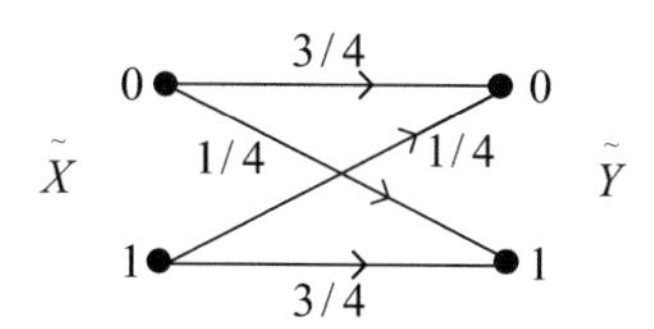

Fig. 2.13 Binary symmetric channel for Example 2.18

Let's explain the capacity for practical applications for the given binary symmetric channel. The capacity $C = 0.1887$ bits/sym means that for repeated transmission of information frames consisting of, for example, 10,000 bits, we can receive 1887 bits correctly in average. This is a guaranteed number. However, this does not mean that for instant transmissions we cannot have more than 1887 bits received correctly. We can have, but, it is totally by change, there is no guarantee for it. We can guarantee the reception of at most 1887 bits in average correctly. This is for sure.

If we generate the bits '0' and '1' at the input of the channel with a distribution other than the uniform distribution, then the mutual information between channel input and output will be smaller than the channel capacity, i.e., will be smaller than the maximum value of the mutual information. In this case, if we transmit frames with 1 10,000 bits through the binary symmetric channel repeatedly, the number of correctly received bits guaranteed will be smaller than the number 1887 in average.

It is very important to understand the meaning of channel capacity. For this purpose, let's study a matlab example to illustrate the meaning of binary symmetric channel capacity.

Example 2.19 For a binary symmetric channel with transmission error probability p the channel capacity, that is the maximum value of the mutual information between channel input and output random variables is achieved when the channel input variable has uniform distribution and in this case the channel capacity is given as $C = 1 - H_b(p)$. Write a matlab code to simulate the channel capacity.

Solution 2.19 First let's calculate the channel capacity of the binary symmetric channel. For this purpose let's choose the transmission error probability $p = 0.008$. The channel capacity calculation can be performed with the matlab code below

```
p=0.008;
C=1+p*log2(p)+(1-p)*log2(1-p)
```

Next, let's simulate the transmission through binary symmetric channel. To obtain the maximum value of the mutual information between channel input and channel output we should generate the channel inputs with uniform distribution. This can be achieved using the matlab code segment below

```
p=0.008;
C=1+p*log2(p)+(1-p)*log2(1-p);
N=75;
x = randi([0 1],1,N); % Random bit vector with uniform distribution
```

In the above code we have chosen bit vector length as $N = 75$. In the next step, let's transmit the generated bit stream through the binary symmetric channel using the matlab function '$bsc(\cdot)$'. Our code happens to be as in

```
p=0.008;
C=1+p*log2(p)+(1-p)*log2(1-p);
N=75;
x = randi([0 1],1,N);  % Random bit vector
y = bsc(x,p); % Binary symmetric channel
num_errs=sum(abs(x-y)); % Number of tranmission errors
cbit_Num=N-num_errs; % Number of correctly received bits
```

After finding the number of correctly received bits, we can calculate the transmission rate as in

```
p=0.008;
C=1+p*log2(p)+(1-p)*log2(1-p);
N=75;
x = randi([0 1],1,N);  % Random bit vector
y = bsc(x,p); % Binary symmetric channel
num_errs=sum(abs(x-y)); % Number of tranmission errors
cbit_Num=N-num_errs; % Number of correctly received bits
```

Now, let's calculate the transmission rate using other data frames, for this purpose, let's use a loop in our program, and modify our program as in

```
clc; clear all;
p=0.008;
C=1+p*log2(p)+(1-p)*log2(1-p);
N=75;
LoopNum=5000;
R_arr=[];
for indx=1:LoopNum
        x = randi([0 1],1,N); % Random matrix
        y = bsc(x,p); % Binary symmetric channel
        numerrs=sum(abs(x-y)); % Number off bit errors
        cbitNum=N-numerrs; % Number of corretly received bits
        R=cbitNum/(N); % Transmision rate
        R_arr=[R_arr R]; % Transmission rate vector
end
plot(R_arr); % Plot the tranmission rate vector
hold on;
plot(1:50:length(R_arr),C,'r.-'); % Draw capacity frontier
legend('R','C')
xlabel('Frame Indx');
ylabel('R');
```

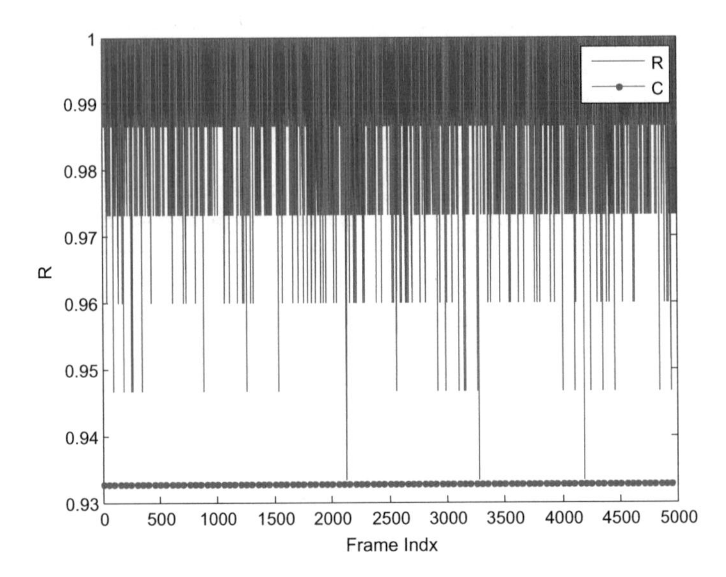

Fig. 2.14 Transmission rate w.r.t. capacity

If we run the above program several times, we can get different graphics, and in most of them the rate graphic will be above the capacity frontier. And in one of the runs, we can get a graph as in Fig. 2.14. This is due to short data vector lengths. Since, we need uniformly distributed information bit vectors, and for short lengths, we can have some bias. In fact, to generate uniformly distributed data in matlab, we should choose the data vector length very large.

As it is seen from Fig. 2.14 that, the rate of the transmission is a random quantity and it is mostly greater than the maximum mutual information by some little amount, i.e., capacity. However, it should not be confused that there is no guarantee that rate is always greater than capacity, it may be or may not be. What is guaranteed in Fig. 2.14 is that we have a guaranteed transmission rate at the channel capacity in average. That is, we are sure that with uniform distribution at the input we have a guaranteed average transmission rate equal to the maximum value of the mutual information. And for better simulation, we can choose the data length N a much more greater number than 75, but, in this case simulation requires much more time.

What happens if the source symbols are not generated uniformly? In that case, the mutual information between channel input and output will be lower than the maximum value of the mutual information which happens when input has uniform distribution. In this case, our guaranteed average transmission rate will be equal to the mutual information value which is smaller than the maximum value of the mutual information.

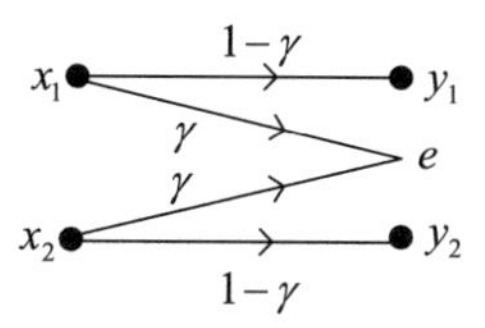

Fig. 2.15 Binary erasure channel for Example 2.20

Capacity of the Binary Erasure Channel

Binary erasure channel is an important channel used in modeling of the communication systems. It is especially used for internet communication. For this reason, we will study the BER capacity computation in details. Let's first solve some problems to prepare ourselves for capacity calculation of binary erasure channel.

Example 2.20 For the binary erasure channel shown in Fig. 2.15, is it possible for the random variable at the channel output to have uniform distribution for any type of distribution available at the input of the channel?

Solution 2.20 Let the input random variable has the distribution

$$p(x_1) = \alpha \quad p(x_2) = 1 - \alpha.$$

The probability mass function of the output random variable can be calculated using

$$p(y) = \sum_x p(x, y) \rightarrow p(y) = \sum_x p(y|x)p(x)$$

as

$$p(y_1) = \alpha(1 - \gamma) \quad p(e) = \gamma \quad p(y_2) = (1 - \alpha)(1 - \gamma).$$

Equating $p(y_1), p(e)$, and $p(y_2)$ to $1/31/3$, we get

$$\alpha(1 - \gamma) = \frac{1}{3} \quad \gamma = \frac{1}{3} \quad (1 - \alpha)(1 - \gamma) = \frac{1}{3}. \tag{2.78}$$

If we divide the equations $\alpha(1 - \gamma) = 1/3$ and $(1 - \alpha)(1 - \gamma) = 1/3$ side by side, we get

$$\frac{\alpha}{1 - \alpha} = 1 \rightarrow \alpha = \frac{1}{2}.$$

However, we see that the output random variable has uniform distribution only when

$$\gamma = \frac{1}{3} \quad \text{and} \quad \alpha = \frac{1}{2}.$$

For any other erasure probability γ different then 1/3, i.e., for

$$\gamma \neq \frac{1}{3}$$

it is not possible to have uniform distribution at the output for any input distribution.

Example 2.21 Calculate the channel capacity of the binary erasure channel given in Fig. 2.16.

Solution 2.21 For the capacity calculation we need the mutual information which can be calculated either using

$$I(\tilde{X};\tilde{Y}) = H(\tilde{Y}) - H(\tilde{Y}|\tilde{X})$$

or using

$$I(\tilde{X};\tilde{Y}) = H(\tilde{X}) - H(\tilde{X}|\tilde{Y}).$$

Both can be utilized. Let's use first use

$$I(\tilde{X};\tilde{Y}) = H(\tilde{X}) - H(\tilde{X}|\tilde{Y}) \tag{2.79}$$

for the capacity calculation. To calculate the entropy $H(\tilde{X})$ in (2.79), we need the input distribution, for this purpose, let $p(x_1) = \alpha$ and $p(x_2) = 1 - \alpha$. The entropy $H(\tilde{X})$ can be calculated as

$$H(\tilde{X}) = -\sum_x p(x)\log p(x) \to H(\tilde{X}) = -[\alpha\log\alpha + (1-\alpha)\log(1-\alpha)] \to$$
$$H(\tilde{X}) = H_b(\alpha). \tag{2.80}$$

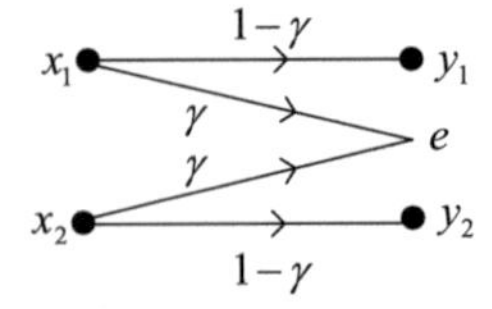

Fig. 2.16 Binary erasure channel for Example 2.21

Next, we can calculate $H(\tilde{X}|\tilde{Y})$ using

$$H(\tilde{X}|\tilde{Y}) = \sum_{y} p(y)H(\tilde{X}|y) \tag{2.81}$$

where

$$H(\tilde{X}|y) = -\sum_{x} p(x|y)\log p(x|y)$$

And for the conditional entropy $H(\tilde{X}|y)$ where y can be y_1, y_2, and e, we have

$$H(\tilde{X}|y_1) = 0 \quad H(\tilde{X}|y_2) = 0$$

since, if y_1 or y_2 is available at the output of the channel, we can know the transmitted symbol without any conflict, i.e., there is no uncertainty about the source $\tilde{X}$. Then, (2.81) reduces to

$$H(\tilde{X}|\tilde{Y}) = p(y=e)H(\tilde{X}|y=e) \tag{2.82}$$

The conditional entropy $H(\tilde{X}|e)$ can be calculated using

$$\begin{aligned} H(\tilde{X}|y=e) &= -\sum_{x} p(x|y=e)\log p(x|y=e) \rightarrow \\ H(\tilde{X}|y=e) &= -[p(x=x_1|y=e)\log p(x=x_1|y=e) \\ &\quad + p(x=x_2|y=e)\log p(x=x_2|y=e)] \end{aligned} \tag{2.83}$$

where the probabilities $p(x=x_1|y=e)$ and $p(x=x_2|y=e)$ can be calculated as

$$p(x=x_1|y=e) = \frac{\underbrace{p(y=e|x=x_1)}_{\gamma}\underbrace{p(x=x_1)}_{\alpha}}{p(y=e)} \rightarrow p(x=x_1|y=e) = \frac{\gamma\alpha}{p(y=e)}$$

$$p(x=x_2|y=e) = \frac{\underbrace{p(y=e|x=x_2)}_{\gamma}\underbrace{p(x=x_2)}_{1-\alpha}}{p(y=e)} \rightarrow p(x=x_2|y=e) = \frac{\gamma(1-\alpha)}{p(y=e)}$$

in which $p(y=e)$ can be calculated as

$$\begin{aligned} &p(y=e) = \sum_{x} p(x, y=e) \rightarrow p(y=e) = \sum_{x} p(y=e|x)p(x) \rightarrow \\ &p(y=e) = p(y=e|x_1)p(x_1) + p(y=e|x_2)p(x_2) \rightarrow \\ &p(y=e) = \gamma\alpha + \gamma(1-\alpha) \rightarrow p(y=e) = \gamma. \end{aligned}$$

Then, we have

$$p(x = x_1|y = e) = \frac{\gamma\alpha}{\gamma} \rightarrow p(x = x_1|y = e) = \alpha$$

$$p(x = x_2|y = e) = \frac{\gamma(1 - \alpha)}{\gamma} \rightarrow p(x = x_2|y = e) = 1 - \alpha$$

Finally, using the computed conditional probabilities in (2.82), we get

$$H(\tilde{X}|\tilde{Y}) = -\gamma[\alpha \log \alpha + (1 - \alpha) \log(1 - \alpha)] \quad (2.84)$$

which can be denoted as

$$H(\tilde{X}|\tilde{Y}) = \gamma H_b(\alpha). \quad (2.85)$$

Now, we are ready to calculate the mutual information. Using (2.85) and (2.80) in (2.79), we obtain

$$I(\tilde{X};\tilde{Y}) = H_b(\alpha) - \beta H_b(\alpha) \rightarrow I(\tilde{X};\tilde{Y}) = (1 - \gamma)H_b(\alpha) \quad (2.86)$$

which is the mutual information in which the source probability α is a variable. Finding capacity is nothing but an optimization problem. We will try to optimize the mutual information considering the variable parameters available in the mutual information expression. The capacity calculation is performed considering

$$C = \max_{p(x)} I(\tilde{X};\tilde{Y}) \rightarrow C = \max(1 - \gamma)H_b(\alpha)$$

leading to

$$C = (1 - \gamma) \max H_b(\alpha)$$

where the binary entropy function $H_b(\alpha)$ gets its maximum value at $\alpha = 1/2$, and its maximum value is $\max H_b(\alpha) = \log 2 = 1$. Hence, capacity of the binary erasure channel becomes equal to

$$C = 1 - \gamma.$$

Solution 2 Now, let's use the alternative mutual information expression

$$I(\tilde{X};\tilde{Y}) = H(\tilde{Y}) - H(\tilde{Y}|\tilde{X}) \quad (2.87)$$

to calculate the channel capacity of the binary erasure channel. The entropy $H(\tilde{Y})$ in (2.87) can be calculated using

$$H(\tilde{Y}) = -\sum_y p(y) \log p(y)$$

where the output probabilities $p(y)$ can be calculated using

$$p(y) = \sum_x p(x, y) \rightarrow p(y) = \sum_x p(y|x)p(x)$$

which can be computed for $y = y_1$, $y = y_2$, and $y = e$ as in

$$p(y = y_1) = (1 - \gamma)\alpha, \quad p(y = y_2) = (1 - \gamma)(1 - \alpha), \quad p(y = e) = \gamma$$

Then, $H(\tilde{Y})$ is calculated as

$$H(\tilde{Y}) = -[(1 - \gamma)\alpha \log((1 - \gamma)\alpha) + (1 - \gamma)(1 - \alpha) \log((1 - \gamma)(1 - \alpha)) + \gamma \log \gamma]$$

which can be, after some manipulation, simplified as

$$H(\tilde{Y}) = H_b(\gamma) + (1 - \gamma)H_b(\alpha)$$

where

$$H_b(\gamma) = -[\gamma \log \gamma + (1 - \gamma) \log(1 - \gamma)] \quad H_b(\alpha) = -[\alpha \log \alpha + (1 - \alpha) \log(1 - \alpha)]$$

The conditional entropy $H(\tilde{Y}|\tilde{X})$ in (2.87) can be calculated using

$$H(\tilde{Y}|\tilde{X}) = \sum_x p(x) H(\tilde{Y}|\tilde{X} = x) \tag{2.88}$$

where

$$H(\tilde{Y}|\tilde{X} = x) = -\sum_y p(y|x) \log p(y|x)$$

which can be computed for $\tilde{X} = x_1$ and $\tilde{X} = x_2$ as

$$H(\tilde{Y}|\tilde{X} = x_1) = -\sum_y p(y|x) \log p(y|x) \rightarrow$$

$$H(\tilde{Y}|\tilde{X} = x_1) = -\left[\underbrace{p(y_1|x_1)}_{1-\gamma} \log \underbrace{p(y_1|x_1)}_{1-\gamma} + \underbrace{p(y_2|x_1)}_{0} \log \underbrace{p(y_2|x_1)}_{0} + \underbrace{p(e|x_1)}_{\gamma} \log \underbrace{p(e|x_1)}_{\gamma}\right]$$

which can be written as

$$H(\tilde{Y}|\tilde{X} = x_1) = H_b(\gamma)$$

In a similar manner, we can get

$$H(\tilde{Y}|\tilde{X} = x_2) = H_b(\gamma).$$

From (2.88), we obtain

$$H(\tilde{Y}|\tilde{X}) = H_b(\gamma).$$

Then, the mutual information expression in (2.87) happens to be

$$I(\tilde{X};\tilde{Y}) = H(\tilde{Y}) - H(\tilde{Y}|\tilde{X}) \to I(\tilde{X};\tilde{Y}) = H_b(\gamma) + (1-\gamma)H_b(\alpha) - H_b(\gamma)$$

which can be written as

$$I(\tilde{X};\tilde{Y}) = (1-\gamma)H_b(\alpha)$$

whose maximum value $(1-\gamma)$ is nothing but the capacity as obtained in the first solution.

Note: For the mutual information $I(\tilde{X};\tilde{Y}) = H(\tilde{Y}) - H(\tilde{Y}|\tilde{X})$, the maximum value of $H(\tilde{Y})$ cannot be equal to $H(\tilde{Y}) = \log|R_{\tilde{Y}}| = \log 3$, since it is not possible to get uniform distribution at the output of the binary erasure channel for any distribution at the input of the channel.

Example 2.22 Calculate the channel capacity of the binary erasure channel given in Fig. 2.17.

Solution 2.22 Using the capacity expression $C = 1-\gamma$, we find the capacity of the binary erasure channel as

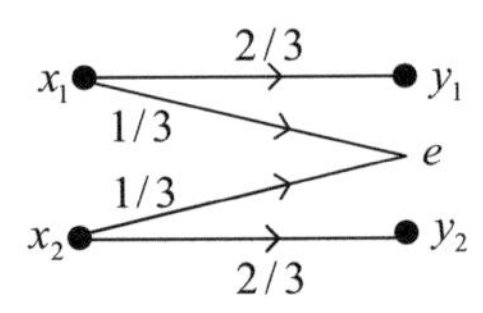

Fig. 2.17 Binary erasure channel for Example 2.22

$$C = 1 - \frac{1}{3} \rightarrow C = \frac{2}{3} \text{ bits/transmission.}$$

Example 2.23 Write a matlab program to simulate the capacity of the binary erasure channel, and take the channel erasure probability $\gamma = 0.6$.

Solution 2.23 While deriving the channel capacity expression for the binary erasure channel, we found that, the channel capacity occurs when the random variable at the input of the binary erasure channel has uniform distribution. For this reason, we need to generate uniformly distributed data, then pass it through binary erasure channel. The following matlab program is written to achieve this goal.

```
clc;clear all;close all;
bec_beta=0.6; % Binary erasure chanel erasure prob.
C=1-bec_beta;
N=1000000;
LoopNum=1000;
R_arr_bec=[]; % Rate array for uniform data simulation
for indx=1:LoopNum
        xbec = randi([0 1],1,N); % Random bit vector
        ybec=xbec; % BEC output
        ybec(rand(size(xbec))<bec_beta)=-1; % Output of the BEC  channel
        numerrs_bec=sum(xbec~=ybec); % Number off bit errors
        cbitNum_bec=N-numerrs_bec; % Number of corretly received bits
        R_bec=cbitNum_bec/(N); % Transmision rate
        R_arr_bec=[R_arr_bec R_bec]; % Transmission rate vector
end
plot(R_arr_bec); % Plot the tranmission rate vector
hold on;
plot(1:50:length(R_arr_bec),C,'r.-'); % Draw capacity frontier
legend('R','C')
xlabel('Frame Indx');
ylabel('R');
```

When the above program is run we obtain the graph shown in Fig. 2.18 where it is seen that the average rate of the transmission is almost equal to capacity of the channel.

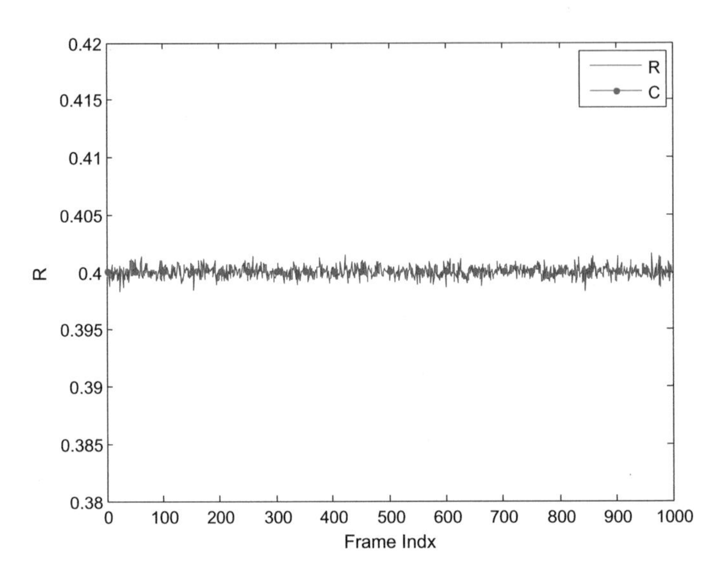

Fig. 2.18 Rate w.r.t capacity

Exercise Calculate the mutual information between input and output of the non-symmetric binary erasure channel depicted in Fig. 2.19.

Example 2.24 For the communication channel shown in Fig. 2.20, if $p(y_1|x_1) = 0.7$, find the other channel transition probabilities.

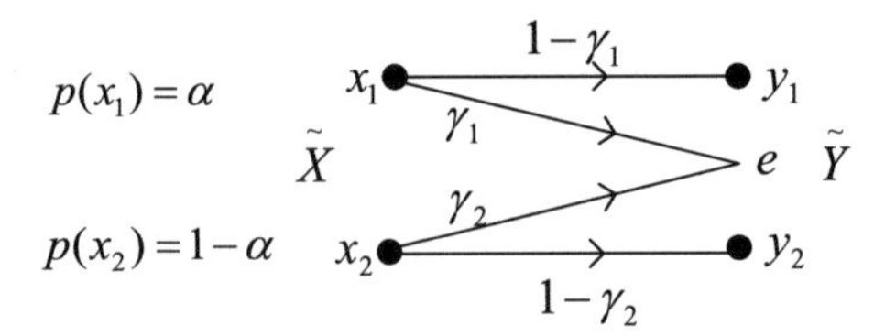

Fig. 2.19 Non-symmetric binary erasure channel

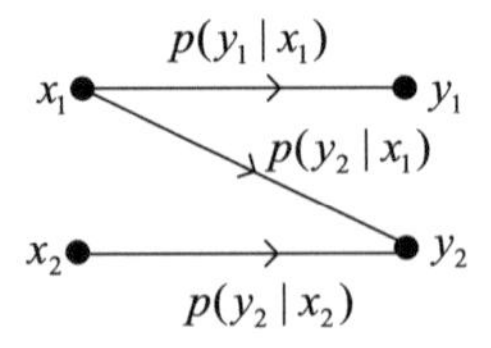

Fig. 2.20 Discrete communication channel for Example 2.24

Solution 2.24 In a discrete memoryless communication channel, we have

$$\sum_{r} p(y_r|x_i) = 1 \tag{2.89}$$

where y_r are the channel outputs, and x_i is one of the channel inputs. We can write (2.89) as

$$\sum_{y} p(y|x) = 1. \tag{2.90}$$

Considering (2.89) or (2.90) for the channel in Fig. 2.20, we can write

$$p(y_1|x_1) + p(y_2|x_1) = 1 \quad p(y_2|x_2) = 1$$

from which, we obtain

$$p(y_2|x_1) = 0.3.$$

Example 2.25 Find the transition probabilities for the channel shown in Fig. 2.21.

Solution 2.25 Using the property

$$\sum_{y} p(y|x) = 1$$

we obtain the equations

$$k + 2k = 1, \quad m + 2m = 1, \quad l = 1$$

from which, we get

$$k = \frac{1}{3}, \quad m = \frac{1}{3}, \quad l = 1.$$

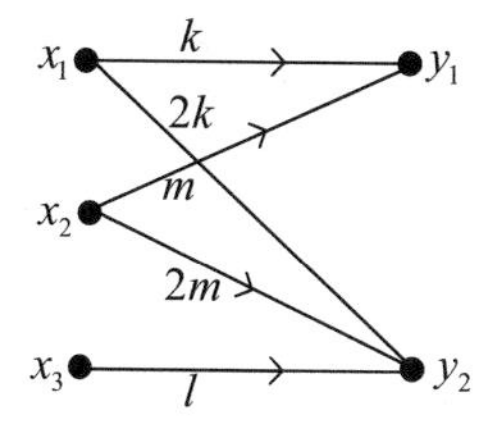

Fig. 2.21 Discrete communication channel for Example 2.25

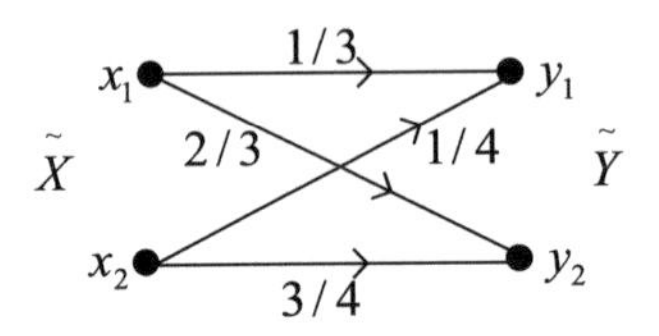

Fig. 2.22 Discrete communication channel for Example 2.26

Example 2.26 For the communication channel depicted in Fig. 2.22, show that output random variable cannot have uniform distribution.

Solution 2.26 If the random variable at the output of the given discrete memoryless channel has uniform distribution, then we have

$$p(y = y_1) = \frac{1}{2} \quad p(y = y_2) = \frac{1}{2}. \tag{2.91}$$

Using the channel transition probabilities, we can write

$$p(y) = \sum_x p(x, y) \rightarrow p(y) = \sum_x p(y|x)p(x)$$

from which, we get

$$\begin{aligned} p(y_1) &= \frac{1}{3}p(x_1) + \frac{1}{4}p(x_2) \\ p(y_2) &= \frac{2}{3}p(x_1) + \frac{3}{4}p(x_2) \end{aligned} \tag{2.92}$$

Using (2.91) and (2.92), we obtain the equation set

$$\begin{aligned} \frac{1}{3}p(x_1) + \frac{1}{4}p(x_2) &= \frac{1}{2} \\ \frac{2}{3}p(x_1) + \frac{3}{4}p(x_2) &= \frac{1}{2} \end{aligned} \tag{2.93}$$

whose solution is

$$p(x_2) = -\frac{2}{3}p(x_1). \tag{2.94}$$

However, the solution in (2.94) has no meaning, since probability cannot be negative. For this reason, we do not have any solution for the equation set in (2.93).

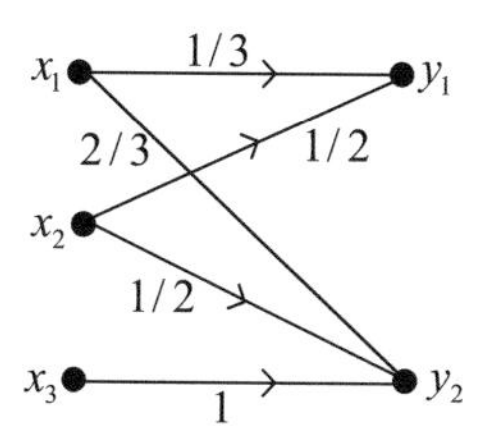

Fig. 2.23 Discrete communication channel for exercise

Exercise For the communication channel show in Fig. 2.23, show that it is not possible to have uniform distribution at the output of the communication channel.

Example 2.27 Find the channel capacity of the discrete memoryless channel shown in Fig. 2.24.

Solution 2.27 Let the input distribution be given as

$$p(x_1) = 1 - \alpha, \quad p(x_2) = \alpha.$$

First, let's calculate the mutual information

$$I(\tilde{X}; \tilde{Y}) = H(\tilde{Y}) - H(\tilde{Y}|\tilde{X}) \tag{2.95}$$

in terms of the variable α, and then try to optimize the mutual information considering the α parameter. The conditional entropy $H(\tilde{Y}|\tilde{X})$ in (2.95) can be calculated using

$$H(\tilde{Y}|\tilde{X}) = \sum_x p(x) H(\tilde{Y}|x)$$

where

$$H(\tilde{Y}|x) = -\sum_y p(y|x) \log p(y|x)$$

which can be calculated for $x = x_1$ and $x = x_2$ as

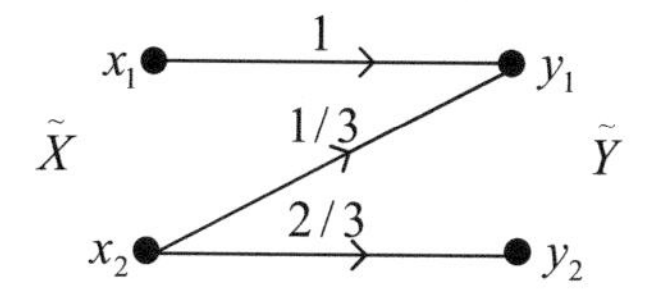

Fig. 2.24 Discrete communication channel for Example 2.27

$$H(\tilde{Y}|x = x_1) = -[1 \log 1] \rightarrow H(\tilde{Y}|x = x_1) = 0$$
$$H(\tilde{Y}|x = x_2) = -\left[\frac{1}{3}\log\frac{1}{3} + \frac{2}{3}\log\frac{2}{3}\right] \rightarrow H(\tilde{Y}|x = x_2) = 0.9183.$$

The conditional entropy $H(\tilde{Y}|\tilde{X})$ can be evaluated as

$$H(\tilde{Y}|\tilde{X}) = \sum_x p(x)H(\tilde{Y}|x) \rightarrow H(\tilde{Y}|\tilde{X}) = 0.9183\alpha. \quad (2.96)$$

The entropy $H(\tilde{Y})$ in (2.95) can be found using

$$H(\tilde{Y}) = -\sum_y p(y)\log p(y) \quad (2.97)$$

where $p(y)$ can be computed for $y = y_1$ and $y = y_2$ as

$$p(y_1) = 1 - \alpha + \frac{1}{3}\alpha \rightarrow p(y_1) = 1 - \frac{2}{3}\alpha$$
$$p(y_2) = 0(1 - \alpha) + \frac{2}{3}\alpha \rightarrow p(y_2) = \frac{2}{3}\alpha$$

Then, $H(\tilde{Y})$ is computed as

$$H(\tilde{Y}) = -\left[\left(1 - \frac{2}{3}\alpha\right)\log\left(1 - \frac{2}{3}\alpha\right) + \left(\frac{2}{3}\alpha\right)\log\left(\frac{2}{3}\alpha\right)\right]. \quad (2.98)$$

Using (2.96) and (2.98), the mutual information $I(\tilde{X};\tilde{Y}) = H(\tilde{Y}) - H(\tilde{Y}|\tilde{X})$ in terms of the variable α is calculated as

$$I(\tilde{X};\tilde{Y}) = -\left[\left(1 - \frac{2}{3}\alpha\right)\log\left(1 - \frac{2}{3}\alpha\right) + \left(\frac{2}{3}\alpha\right)\log\left(\frac{2}{3}\alpha\right)\right] - 0.9183\alpha. \quad (2.99)$$

Capacity is nothing but the maximum value of the mutual information in (2.99), i.e.

$$C = \max_{p(x)} I(\tilde{X};\tilde{Y}).$$

To find the maximum value of the mutual information, we take the derivative of (2.99) w.r.t. α and equate it to zero as in

$$\frac{\partial I(\tilde{X};\tilde{Y})}{\partial\alpha} = 0 \rightarrow -\left[-\frac{2}{3}\log\left(1 - \frac{2}{3}\alpha\right) + \left(\frac{2}{3}\right)\log\left(\frac{2}{3}\alpha\right)\right] - 0.9183 = 0$$

from which, we get

$$\frac{2}{3}\log\left(\frac{1-\frac{2}{3}\alpha}{\frac{2}{3}\alpha}\right)=0.9183$$

whose solution is

$$\alpha = 0.4169$$

which can be used to calculate the maximum value of the mutual information in (2.99) as

$$\max I\left(\tilde{X};\,\tilde{Y}\right) = 0.4968 \text{ bits/transmission}$$

which is nothing but the capacity of the channel given in the question, i.e.,

$$C = 0.4968 \text{ bits/transmission.}$$

Exercise For the discrete memoryless channel shown in Fig. 2.25, the transition probability p is a constant number. Calculate the channel capacity of the channel in Fig. 2.25.

Example 2.28 For the communication channel shown in Fig. 2.26, the input random variables $\tilde{X}$ and $\tilde{Z}$ are uniformly distributed and their range sets are $R_{\tilde{X}} = \{0, 1, 2, 3, 4\}$, $R_{\tilde{Z}} = \{1, 3\}$. The output of the channel is expressed as $\tilde{Y} = (\tilde{X} + \tilde{Z}) \bmod 5$.

Find the channel outputs, and decide on the distribution at the channel output. Find the channel transition probabilities, and redraw the channel using its transition probabilities.

Solution 2.28 We can calculate the channel outputs using $y = x + z$ where x is a single integer chosen from $R_{\tilde{X}}$, and z can be considered as noise sample taking

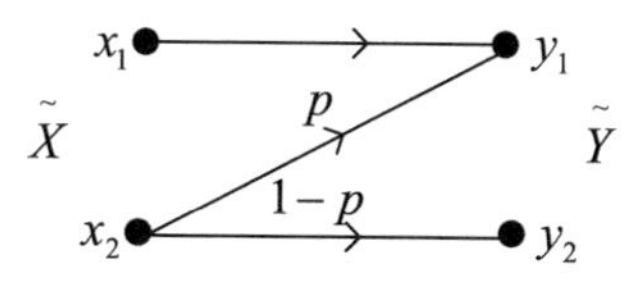

Fig. 2.25 Discrete communication channel for exercise

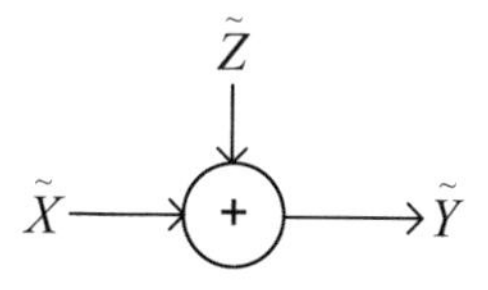

Fig. 2.26 Communication channel for Example 2.28

Table 2.1 Channel outputs

$\tilde{X}$	$\tilde{Y}$
0	1.3
1	2.4
2	3.0
3	4.1
4	0.2

values 1 and 3 respectively. For instance, for $x = 1$, y can be calculated as $y = 1 + 1 \ mod \ 5 \rightarrow y = 2$, or $y = 1 + 3 \ mod \ 5 \rightarrow y = 4$. Hence, for $x = 1$ y can be 2 or 4. Considering all input values, the channel outputs can be computed as in Table 2.1.

Table 2.1 can also be expressed via channel transition probabilities as in Fig. 2.27 where all the transition probabilities are 1/2, i.e., $p(y|x) = 1/2$, for $\forall x, y$.

Since the input random variable is uniformly distributed, we have $p(x) = 1/5$ for $x = 0, 1, 2, 3, 4$. The output distribution can be calculated using

$$p(y) = \sum_x p(x, y) \rightarrow p(y) = \sum_x p(x)p(y|x)$$

and it can be shown that output random variable is also uniformly distributed, i.e., $p(y) = 1/5$ for $y = 0, 1, 2, 3, 4$.

Example 2.29 For the discrete random variables $\tilde{X}$, $\tilde{Y}$ and $\tilde{Z}$, if $\tilde{Y} = \tilde{X} + \tilde{Z}$, show that

$$H(\tilde{Y}|\tilde{X}) = H(\tilde{Z}|\tilde{X}).$$

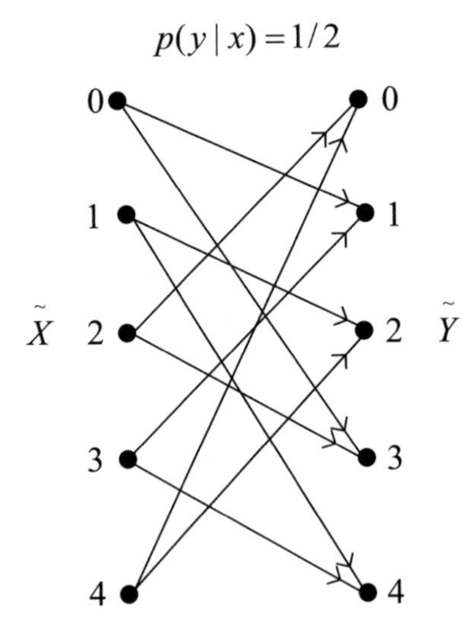

Fig. 2.27 Equivalent channel graph

Solution 2.29 The conditional entropy $H(\tilde{Y}|\tilde{X})$ can be calculated as

$$H(\tilde{Y}|\tilde{X}) = -\sum_{x,y} p(x,y)\log p(y|x) \rightarrow H(\tilde{Y}|\tilde{X}) = -\sum_{x,y} p(x)p(y|x)\log p(y|x)$$

where the conditional probability $p(y|x)$ can be shown to be equal to $p(z|x)$ as follows

$$p(y|x) = Prob(\tilde{Y} = y|\tilde{X} = x) \rightarrow p(y|x) = Prob(\tilde{Y} = x + z|\tilde{X} = x)$$
$$Prob(\tilde{Y} = x + z|\tilde{X} = x) = Prob(\tilde{Z} = z|\tilde{X} = x) \rightarrow p(y|x) = p(z|x).$$

Then, we have

$$H(\tilde{Y}|\tilde{X}) = -\sum_{x,y} p(x)\underbrace{p(y|x)}_{p(z|x)}\log \underbrace{p(y|x)}_{p(z|x)} \rightarrow H(\tilde{Y}|\tilde{X}) = -\sum_{x,z} p(x)p(z|x)\log p(z|x)$$

Hence, we get

$$H(\tilde{Y}|\tilde{X}) = H(\tilde{Z}|\tilde{X}).$$

Note: If $\tilde{Z}$ and $\tilde{X}$ are independent discrete random variables, then we have

$$H(\tilde{Z}|\tilde{X}) = H(\tilde{Z}).$$

Example 2.30 For the communication channel shown in Fig. 2.28, the input random variables $\tilde{X}$ and $\tilde{Z}$ are uniformly distributed and their range sets are $R_{\tilde{X}} = \{0, 1, 2, 3, 4\}$, $R_{\tilde{Z}} = \{1, 3\}$. The output of the channel is expressed as $\tilde{Y} = (\tilde{X} + \tilde{Z})\ mod\ 5$. Calculate the capacity between the random variables $\tilde{X}$ and $\tilde{Y}$, i.e., calculate the channel capacity.

Solution 2.30 The mutual information between $\tilde{X}$ and $\tilde{Y}$ can be expressed as

$$I(\tilde{X}, \tilde{Y}) = H(\tilde{Y}) - H(\tilde{Y}|\tilde{X}). \tag{2.100}$$

In our previous two examples, we have seen that $H(\tilde{Y}|\tilde{X}) = H(\tilde{Z}|\tilde{X})$, and since the random variables $\tilde{X}$ and $\tilde{Z}$ are independent of each other, we have

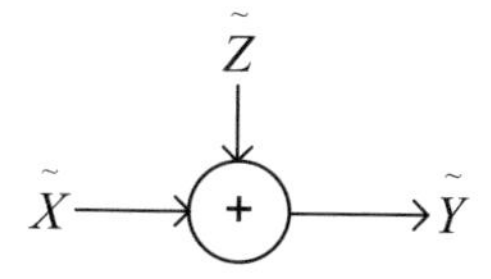

Fig. 2.28 Communication channel for Example 2.30

$H(\tilde{Y}|\tilde{X}) = H(\tilde{Z}|\tilde{X}) = H(\tilde{Z})$. Then, the mutual information expression in (2.100) happens to be as in

$$I(\tilde{X}, \tilde{Y}) = H(\tilde{Y}) - H(\tilde{Z})$$

where $H(\tilde{Y}) = \log|R_{\tilde{Y}}| \rightarrow H(\tilde{Y}) = \log 5$, and $H(\tilde{Z}) = \log|R_{\tilde{Z}}| \rightarrow H(\tilde{Z}) = \log 2$, since both $\tilde{Y}$ and $\tilde{Z}$ are uniformly distributed. Thus, the capacity of the channel can be written as

$$C = \log\frac{5}{2} \text{ bits/tranmission.}$$

Exercise For the previous example, if the number of elements in the range set $R_{\tilde{Z}}$ goes to infinity, what happens to the channel capacity?

Example 2.31 Calculate the capacity of the discrete memoryless channel shown in Fig. 2.29 where $p(x_1) = \alpha, p(x_2) = \alpha$ and $p(x_3) = 1 - 2\alpha$.

Solution 2.31 Let's first calculate the mutual information

$$I(\tilde{X}, \tilde{Y}) = H(\tilde{Y}) - H(\tilde{Y}|\tilde{X}) \tag{2.101}$$

where for the computation of $H(\tilde{Y})$, we need the output symbol probabilities which can be calculated using

$$p(y) = \sum_x p(x, y) \rightarrow p(y) = \sum_x p(x)p(y|x)$$

as in

$$p(y_1) = \alpha, \quad p(y_2) = \alpha, \quad p(y_3) = 1 - 2\alpha.$$

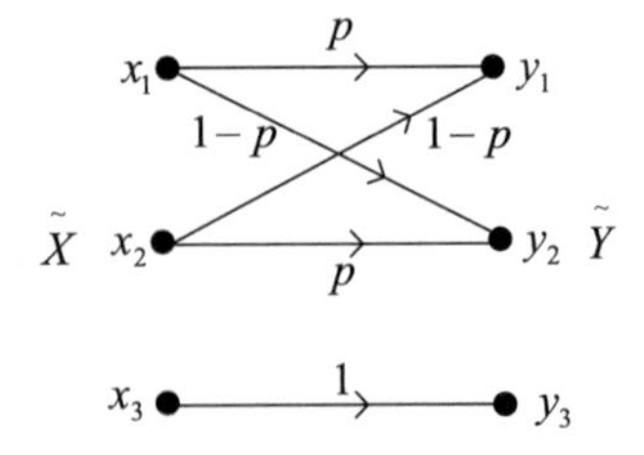

Fig. 2.29 Discrete communication channel for Example 2.31

Then, the entropy $H(\tilde{Y})$ is determined as

$$H(\tilde{Y}) = -\sum_y p(y)\log p(y) \rightarrow H(\tilde{Y}) = -[\alpha\log\alpha + \alpha\log\alpha + (1-2\alpha)\log(1-2\alpha)] \rightarrow$$
$$H(\tilde{Y}) = -[2\alpha\log\alpha + (1-2\alpha)\log(1-2\alpha)].$$

The conditional mutual information $H(\tilde{Y}|\tilde{X})$ can be calculated using

$$H(\tilde{Y}|\tilde{X}) = \sum_x p(x)H(\tilde{Y}|x)$$

where $H(\tilde{Y}|x)$ for x_1, x_2, and x_3 can be evaluated using

$$H(\tilde{Y}|x) = -\sum_y p(y|x)\log p(y|x)$$

as

$$H(\tilde{Y}|x_1) = -[p\log p + (1-p)\log(1-p)] \rightarrow H(\tilde{Y}|x_1) = H_b(p)$$
$$H(\tilde{Y}|x_2) = -[p\log p + (1-p)\log(1-p)] \rightarrow H(\tilde{Y}|x_1) = H_b(p)$$
$$H(\tilde{Y}|x_3) = 0.$$

Then, the conditional mutual information can be calculated as

$$H(\tilde{Y}|\tilde{X}) = 2\alpha H_b(p).$$

Now, we can evaluate the mutual information in (2.101) as

$$I(\tilde{X},\tilde{Y}) = -2\alpha\log\alpha - (1-2\alpha)\log(1-2\alpha) - 2\alpha H_b(p). \tag{2.102}$$

If we take the derivative of $I(\tilde{X},\tilde{Y})$ w.r.t $\alpha\alpha$ and equate it to zero, i.e.,

$$\frac{\partial I(\tilde{X},\tilde{Y})}{\partial\alpha} = 0$$

we get

$$H_b(p) = \log(\alpha^{-1} - 2). \tag{2.103}$$

If (2.103) is substituted into (2.102), we obtain

$$I(\tilde{X},\tilde{Y}) = -2a\log\alpha - (1-2\alpha)\log(1-2\alpha) - 2\alpha\log\left(\alpha^{-1}-2\right) \tag{2.104}$$

where the first term and the last term on the right side can be combined as

$$I(\tilde{X},\tilde{Y}) = -2\alpha\left(\log\alpha + \log\left(\alpha^{-1}-2\right)\right) - (1-2\alpha)\log(1-2\alpha)$$

leading to

$$I(\tilde{X},\tilde{Y}) = -2\alpha\left(\log\left(\alpha\left(\alpha^{-1}-2\right)\right)\right) - (1-2\alpha)\log(1-2\alpha)$$

which can be simplified as

$$I(\tilde{X},\tilde{Y}) = -2\alpha\left(\log\left(\alpha\left(\alpha^{-1}-2\right)\right)\right) - (1-2\alpha)\log(1-2\alpha) \tag{2.105}$$

where α can be expressed in terms of $H_b(p)$ by solving (2.103) as

$$\alpha = \left(2 + 2^{H_b(p)}\right)^{-1}.$$

Thus, the capacity of the channel can be expressed as

$$C = -\log\left(1 - 2\left(2 + 2^{H_b(p)}\right)^{-1}\right).$$

Note:

$$\frac{\partial \log_a f(x)}{\partial x} = \frac{f'(x)}{f(x)}\log_\alpha e$$

Example 2.32 The noisy typewriter channel is shown in Fig. 2.30 where all the channel transition probabilities are equal to 1/2, i.e., $p(y|x) = 1/2, \forall x, y$. The input symbols are uniformly distributed, i.e., $p(A) = p(B) = \cdots = p(Z) = 1/26$, i.e., the input random variable is uniformly distributed. Find the distribution of the output symbols, i.e., find the distribution of the output random variable.

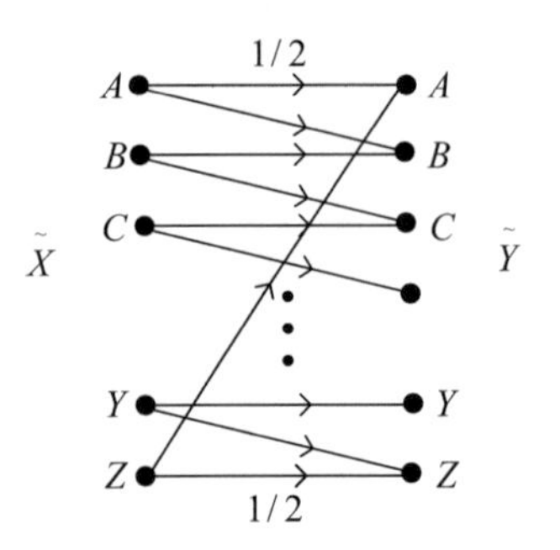

Fig. 2.30 The noisy typewriter channel

Solution 2.32 The probability mass function of the output random variable can be calculated using

$$p(y) = \sum_x p(x,y) \rightarrow p(y) = \sum_x p(y|x)p(x). \tag{2.106}$$

For instance, (2.106) can be calculated for $y = A$ as

$$p(y = A) = p(y = A|x = A)p(x = A) + p(y = A|x = Z)p(x = Z)$$

leading to

$$p(y = A) = \frac{1}{2} \times \frac{1}{26} + \frac{1}{2} \times \frac{1}{26} \rightarrow p(y = A) = \frac{1}{26}.$$

If (2.106) is computed for other output symbols, we see that

$$p(y = B) = p(y = C) = \cdots = p(y = Z) = \frac{1}{26}.$$

That is, the output random variable is also uniformly distributed.

Example 2.33 The noisy typewriter channel is shown in Fig. 2.31 where all the channel transition probabilities are equal to 1/2, i.e., $p(y|x) = 1/2, \forall\, x, y$. The noisy typewriter channel can be interpreted as follows. Assume that you want to press the key 'A', this occurs with probability 1/2, or you press the following key with the same probability, i.e., 1/2. The input symbols are uniformly distributed, i.e., $p(A) = p(B) = \cdots = p(Z) = 1/26$.
Find the capacity of the noisy typewriter channel.

Solution 2.33 To find the channel capacity, let's first calculate the mutual information

$$I(\tilde{X}; \tilde{Y}) = H(\tilde{X}) - H(\tilde{X}|\tilde{Y}). \tag{2.107}$$

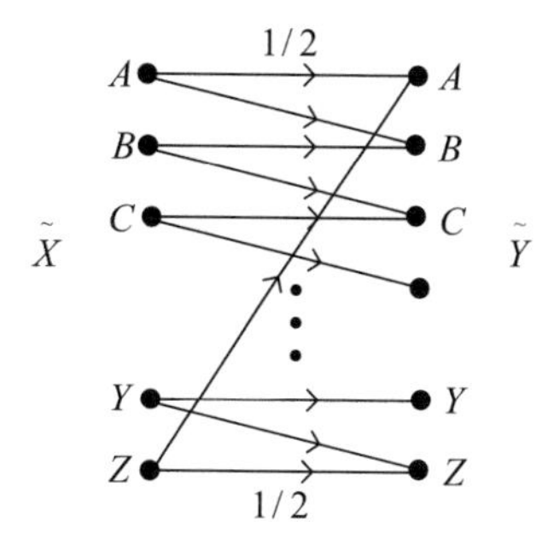

Fig. 2.31 The noisy typewriter channel for Example 2.33

The input random variable has the range set $R_{\tilde{X}}$ consisting of 26 symbols, i.e., $R_{\tilde{X}} = \{A, B, C, \ldots, Z\}$ and $|R_{\tilde{X}}| = 26$. Since the input random variable has uniform distribution, then we have

$$H(\tilde{X}) = \log|R_{\tilde{X}}| \rightarrow H(\tilde{X}) = \log 26. \tag{2.108}$$

We can calculate the probabilities of the output symbols using

$$p(y) = \sum_x p(x, y) \rightarrow p(y) = \sum_x p(y|x)p(x) \rightarrow p(y) = \frac{1}{26}$$

where $y \in R_{\tilde{Y}} = \{A, B, C, \ldots, Z\}$. That is, the output random is also uniformly distributed. The conditional probability $p(x|y)$ can be calculated as

$$p(x|y) = \frac{p(x, y)}{p(y)} \rightarrow p(x|y) = \frac{\overbrace{p(y|x)}^{1/2}\overbrace{p(x)}^{1/26}}{\underbrace{p(y)}_{1/26}} \rightarrow p(x|y) = \frac{1}{2}. \tag{2.109}$$

The conditional entropy $H(\tilde{X}|\tilde{Y})$ can be computed as

$$H(\tilde{X}|\tilde{Y}) = -\sum_{x,y} p(x, y) \log p(x|y)$$

in which substituting (2.109) for $p(x|y)$, we obtain

$$H(\tilde{X}|\tilde{Y}) = -\sum_{x,y} p(x, y) \log \frac{1}{2} \rightarrow H(\tilde{X}|\tilde{Y}) = \sum_{x,y} p(x, y) \rightarrow H(\tilde{X}|\tilde{Y}) = 1. \tag{2.110}$$

Using (2.108) and (2.110) in (2.107), we get

$$I(\tilde{X}; \tilde{Y}) = \log 26 - 1 \rightarrow I(\tilde{X}; \tilde{Y}) = \log 26 - \log 2 \rightarrow I(\tilde{X}; \tilde{Y}) = \log 13.$$

Thus, we found that the mutual information is a constant quantity, i.e., $I(\tilde{X}; \tilde{Y}) = \log 13$. This means that its maximum value equals to itself, i.e.,

$$C = I(\tilde{X}; \tilde{Y}) \rightarrow C = \log 13.$$

If we had used

$$I(\tilde{X}; \tilde{Y}) = H(\tilde{Y}) - H(\tilde{Y}|\tilde{X})$$

for the calculation of the mutual information, we would get the same result in an easier manner. Since,

$$H(\tilde{Y}|\tilde{X}) = -\sum_{x,y} p(x,y) \log \underbrace{p(y|x)}_{\frac{1}{2}} \rightarrow H(\tilde{Y}|\tilde{X}) = 1$$

and we do not need the calculation of $p(x|y)$.

Exercise For the noisy type write channel, assume that, we have a transition from every input letter to the same and two consecutive output letters with transition probability $p(y|x) = 1/3$. Find the channel capacity.

2.4 Capacity for Continuous Channels, i.e., Continuous Random Variables

Let $I(\tilde{X}; \tilde{Y})$ be the mutual information between the continuous random variables $\tilde{X}$ and $\tilde{Y}$. Since the range sets of $\tilde{X}$ and $\tilde{Y}$ include uncountable number of elements it is not possible to show all the transition probabilities between the values of $\tilde{X}$ and $\tilde{Y}$. However, we can graphically show it as in Fig. 2.32 where x and y are the values generated by the random variables $\tilde{X}$ and $\tilde{Y}$.

Example 2.34 Let the continuous random variables $\tilde{X}$ and $\tilde{Y}$ be related to each other as $\tilde{Y} = \tilde{X} + \tilde{N}$ where $\tilde{N} \sim N(0, \sigma^2)$, i.e.,

$$p_{\tilde{N}}(n) = \frac{1}{\sqrt{2\pi\sigma^2}} e^{-\frac{n}{2\sigma^2}}.$$

The conditional probability density function $p(y|x)$, i.e., channel transition probabilities, can be expressed as

$$p(y|x) = p_{\tilde{N}}(y - x) \rightarrow p(y|x) = \frac{1}{\sqrt{2\pi\sigma^2}} e^{-\frac{(y-x)^2}{2\sigma^2}}.$$

Example 2.35 Let the continuous random variables $\tilde{X}$ and $\tilde{Y}$ be related to each other $\tilde{Y} = \tilde{X} + \tilde{N}$. The random variable $\tilde{N}$ is independent from $\tilde{X}$ and $\tilde{Y}$. Show that

$$h(\tilde{Y}|\tilde{X}) = h(\tilde{N}).$$

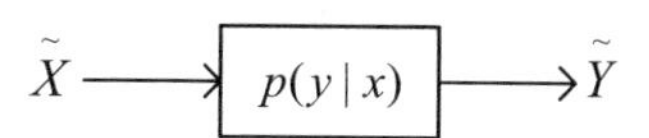

Fig. 2.32 Continuous channel

Solution 2.35 Using the definition of the conditional entropy

$$h(\tilde{Y}|\tilde{X}) = \int f(y|x)\log f(x,y)dxdy,$$

it can be shown that

$$h(\tilde{Y}|\tilde{X}) = h(\tilde{X}+\tilde{N}|\tilde{X}) \rightarrow h(\tilde{Y}|\tilde{X}) = h(\tilde{N}|\tilde{X}) \rightarrow h(\tilde{Y}|\tilde{X}) = h(\tilde{N}).$$

Example 2.36 Let the continuous random variables $\tilde{X}$ and $\tilde{Y}$ be related to each other $\tilde{Y} = \tilde{X}+\tilde{N}$ where $\tilde{N} \sim N(0,\sigma^2)$, i.e.,

$$p_{\tilde{N}}(n) = \frac{1}{\sqrt{2\pi\sigma^2}} e^{-\frac{n^2}{2\sigma^2}}.$$

The random variable $\tilde{N}$ is independent from $\tilde{X}$ and $\tilde{Y}$. Find $h(\tilde{Y}|\tilde{X})$.

Solution 2.36 We know that the entropy of the zero mean Gaussian random variable $\tilde{N}$ with variance $\sigma^2_{\tilde{N}}$ equals to

$$h(\tilde{N}) = \frac{1}{2}\log(2\pi e\sigma^2_{\tilde{N}}).$$

And in the previous example, we found that

$$h(\tilde{Y}|\tilde{X}) = h(\tilde{N}).$$

Hence, we have

$$h(\tilde{Y}|\tilde{X}) = \frac{1}{2}\log(2\pi e\sigma^2_{\tilde{N}})$$

i.e.,

$$h(\tilde{Y}|\tilde{X}) = \frac{1}{2}\log(2\pi e \times Var(\tilde{N})).$$

Example 2.37 Let the continuous random variables $\tilde{X}$ and $\tilde{Y}$ be related to each other as

$$\tilde{Y} = \tilde{X}+\tilde{N}.$$

The random variable $\tilde{N}$ is independent from $\tilde{X}$ and $\tilde{Y}$. Show that

$$Var(\tilde{Y}) = Var(\tilde{X}) + Var(\tilde{N})$$

i.e.,

$$\sigma_{\tilde{Y}}^2 = \sigma_{\tilde{X}}^2 + \sigma_{\tilde{N}}^2.$$

Solution 2.37 The variance of the random variable $\tilde{Y}$ can be calculated as

$$Var(\tilde{Y}) = E\left[(\tilde{X}+\tilde{N})^2\right] - \left[E(\tilde{X}+\tilde{N})\right]^2 \tag{2.111}$$

where $E\left[(\tilde{X}+\tilde{N})^2\right]$ and $\left[E(\tilde{X}+\tilde{N})\right]^2$ can be expanded as

$$\begin{aligned} E\left[(\tilde{X}+\tilde{N})^2\right] &= E[\tilde{X}^2 + \tilde{N}^2 + 2\tilde{X}\tilde{N}] \rightarrow \\ E\left[(\tilde{X}+\tilde{N})^2\right] &= E[\tilde{X}^2] + E[\tilde{N}^2] + 2E[\tilde{X}]E[\tilde{N}] \end{aligned} \tag{2.112}$$

$$\begin{aligned} \left[E(\tilde{X}+\tilde{N})\right]^2 &= \left[E(\tilde{X}) + E(\tilde{N})\right]^2 \rightarrow \\ \left[E(\tilde{X}+\tilde{N})\right]^2 &= \left[E(\tilde{X})\right]^2 + \left[E(\tilde{N})\right]^2 + 2E[\tilde{X}]E[\tilde{N}]. \end{aligned} \tag{2.113}$$

Substituting the results of (2.112) and (2.113) into (2.11), we obtain

$$Var(\tilde{Y}) = \underbrace{E[\tilde{X}^2] - \left[E(\tilde{X})\right]^2}_{Var(\tilde{X})} + \underbrace{E[\tilde{N}^2] - \left[E(\tilde{N})\right]^2}_{Var(\tilde{N})} + \underbrace{2E[\tilde{X}]E[\tilde{N}] - 2E[\tilde{X}]E[\tilde{N}]}_{=0}$$

$$\rightarrow$$

$$Var(\tilde{Y}) = Var(\tilde{X}) + Var(\tilde{N})$$

Example 2.38 Show that the sum of two independent Gaussian distributed random variables yields another Gaussian distributed random variable.

Solution 2.38 Let the independent continuous random variables $\tilde{X}$ and $\tilde{Y}$ have the distribution functions $f(x), g(y)$, and if $\tilde{Z} = \tilde{X} + \tilde{Y}$, then the probability density function of $\tilde{Z}$ equals to the convolution of $f(x)$ and $g(y)$, i.e.,

$$f(z) = \int_{-\infty}^{\infty} f(x)g(z-x)dx. \tag{2.114}$$

For the Gaussian distributed random variables

$$\tilde{X} \sim N(\mu_1, \sigma_1^2) \quad \tilde{Y} \sim N(\mu_2, \sigma_2^2)$$

if $\tilde{Z} = \tilde{X} + \tilde{Y}$ then it can be shown that

$$f(z) = \int_{-\infty}^{\infty} f(x)g(z-x)dx$$

leads to

$$\tilde{Z} \sim N(\mu_1 + \mu_2, \sigma_1^2 + \sigma_2^2).$$

For the simplicity of the proof, let's assume that

$$\tilde{X} \sim N(0,1) \quad \tilde{Y} \sim N(0, \sigma_2^2) \tag{2.115}$$

and employing (2.114) for (2.115), we get

$$f(z) = \int_{-\infty}^{\infty} \frac{1}{\sqrt{2\pi}} e^{-\frac{x^2}{2}} \frac{1}{\sqrt{2\pi}} e^{-\frac{(z-x)^2}{2}} dx$$

which can be written as

$$f(z) = \frac{e^{-z^2/4}}{2\pi} \int_{-\infty}^{\infty} e^{-x^2} e^{zx} e^{-z^2/4} dx$$

leading to

$$f(z) = \frac{e^{-z^2/4}}{\sqrt{4\pi}} \times \underbrace{\frac{1}{\sqrt{\pi}} \int_{-\infty}^{\infty} e^{-(x-z/2)^2} dx}_{=1}$$

where

$$\frac{1}{\sqrt{\pi}} \int_{-\infty}^{\infty} e^{-(x-z/2)^2} dx = 1$$

since

$$\frac{1}{\sqrt{\pi}}\mathrm{e}^{-(x-z/2)^2} \to N\left(\frac{z}{2},\frac{1}{\sqrt{2}}\right).$$

Thus,

$$f(z) = \frac{\mathrm{e}^{-z^2/4}}{\sqrt{4\pi}}$$

which is nothing but a Gaussian distribution, and the random variable $\tilde{Z}$ can be indicated as

$$\tilde{Z} \sim N(0,2).$$

Exercise Show that the sum of a Gaussian and a non-Gaussian distributed independent random variables does not yield a Gaussian distributed random variable. For instance, consider the sum of a uniform and a Gaussian random variable.

Exercise Two exponentially distributed independent continuous random variables are summed. Find the distribution of the resulting random variable. The exponential distribution is characterized by $f(x) = \lambda \mathrm{e}^{-\lambda x},\ x \geq 0$.

Example 2.39 Let the continuous random variables $\tilde{X}$ and $\tilde{Y}$ be related to each other by $\tilde{Y} = \tilde{X} + \tilde{N}$ where $\tilde{N} \sim N(0,\sigma^2)$, i.e.,

$$p_{\tilde{N}}(n) = \frac{1}{\sqrt{2\pi\sigma^2}}\mathrm{e}^{-\frac{n^2}{2\sigma^2}}.$$

Calculate the maximum value of the mutual information

$$I\left(\tilde{X};\tilde{Y}\right) = H\left(\tilde{Y}\right) - H\left(\tilde{Y}|\tilde{X}\right) \tag{2.116}$$

and decide on the distribution of $\tilde{X}$ for which mutual information expression gets its maximum value.

Solution 2.39 From Example 2.33, we have seen that

$$H\left(\tilde{Y}|\tilde{X}\right) = H\left(\tilde{N}\right) \to H\left(\tilde{Y}|\tilde{X}\right) = \frac{1}{2}\log\left(2\pi \times Var\left(\tilde{N}\right)\right). \tag{2.117}$$

Then, the mutual information expression in (2.116) becomes as

$$I(\tilde{X};\tilde{Y}) = H(\tilde{Y}) - \frac{1}{2}\log(2\pi \times Var(\tilde{N})) \tag{2.118}$$

which takes its maximum value if $H(\tilde{Y})$ gets its maximum value. $H(\tilde{Y})$ is the entropy of a continuous random variable, and $H(\tilde{Y})$ gets its maximum value if the random variable $\tilde{Y}$ is Gaussian distributed. In this case, the mutual information expression in (2.118) can be written as

$$\begin{aligned} I(\tilde{X};\tilde{Y}) &= \frac{1}{2}\log(2\pi \times Var(\tilde{Y})) - \frac{1}{2}\log(2\pi \times Var(\tilde{N})) \rightarrow \\ I(\tilde{X};\tilde{Y}) &= \frac{1}{2}\log\left(\frac{Var(\tilde{Y})}{Var(\tilde{N})}\right) \end{aligned} \tag{2.119}$$

The resulting mutual information expression in (2.119) is obtained if $\tilde{Y}$ has Gaussian distribution. This is possible if $\tilde{X}$ is also Gaussian distributed. Since

$$\tilde{Y} = \tilde{X} + \tilde{N}$$

where $\tilde{N}$ is Gaussian distributed, then $\tilde{X}$ should also be Gaussian distributed. Since we know that the sum of two Gaussian distributed random variables yields another Gaussian distributed random variable.
Hence, we can say that the maximum value of the mutual information in (2.119) is achieved if $\tilde{X}$ has Gaussian distribution, which results in Gaussian distributed $\tilde{Y}$.

Example 2.40 Let the continuous random variables $\tilde{X}_1$, $\tilde{X}_2$ and $\tilde{Y}_1$, $\tilde{Y}_2$ be related to each other as

$$\tilde{Y}_1 = \tilde{X}_1 + \tilde{N} \quad \tilde{Y}_2 = \tilde{X}_2 + \tilde{N}$$

where $\tilde{N} \sim N(0, \sigma^2)$. The random variable $\tilde{X}_1$ is Gaussian distributed, whereas the random variable $\tilde{X}_2$ is Laplacian distributed. Compare the magnitude of the mutual information$I(\tilde{X}_1, \tilde{Y}_1)$ and $I(\tilde{X}_2, \tilde{Y}_2)$ to each other.

Solution 2.40 It is obvious that

$$I(\tilde{X}_1, \tilde{Y}_1) > I(\tilde{X}_2, \tilde{Y}_2).$$

Since, mutual information takes its peak value if the random variable added to $\tilde{N}$ is also Gaussian distributed.

Exercise The continuous independent random variables $\tilde{X}$ and $\tilde{N}$ are uniformly distributed on $0 \leq x \leq 1$ and $0 \leq n \leq 1$, i.e., their distributions $f(x)$ and $g(n)$ can be written as

$$f(x) = \begin{cases} 1 & 0 \leq x \leq 1 \\ 0 & otherwise, \end{cases} \quad g(n) = \begin{cases} 1 & 0 \leq n \leq 1 \\ 0 & otherwise. \end{cases} \tag{2.120}$$

Let $\tilde{Y} = \tilde{X} + \tilde{N}$, the probability density function, $k(y)$, of $\tilde{Y}$ can be calculated via

$$k(y) = \int_{-\infty}^{\infty} f(x)g(y-x)dx$$

leading to

$$k(y) = \begin{cases} y & if 0 \leq y \leq 1 \\ 2-y & if 1 \leq y \leq 2 \\ 0 & otherwise. \end{cases}$$

Calculate the mutual information between $\tilde{X}$ and $\tilde{Y}$, i.e., $I(\tilde{X}; \tilde{Y}) = ?$

Exercise The continuous independent random variables $\tilde{X}$ and $\tilde{X}$ are exponentially distributed and their distributions $\tilde{X}$ and $g(n)$ can be written as

$$f(x) = \begin{cases} \lambda e^{-\lambda x} & x \geq 0 \\ 0 & otherwise, \end{cases} \quad g(n) = \begin{cases} \lambda e^{-\lambda n} & n \geq 0 \\ 0 & otherwise. \end{cases}$$

Let $\tilde{Y} = \tilde{X} + \tilde{N}$, the probability density function, $k(y)$, of $\tilde{Y}$ can be calculated via

$$k(y) = \int_{-\infty}^{\infty} f(x)g(y-x)dx$$

leading to

$$k(y) = \left\{ \begin{array}{cc} \lambda^2 y e^{-\lambda y} & y \geq 0 \\ 0 & otherwise. \end{array} \right\}$$

Calculate the mutual information between $\tilde{X}$ and $\tilde{Y}$, i.e., $I(\tilde{X}; \tilde{Y}) = ?$

2.4.1 Capacity of the Gaussian Channel with Power Constraint

In our previous section, we found that if $\tilde{Y} = \tilde{X} + \tilde{N}$ where $\tilde{X} \sim N(\mu_{\tilde{X}}, \sigma^2_{\tilde{X}})$, $\tilde{N} \sim N(\mu_{\tilde{N}}, \sigma^2_{\tilde{N}})$, $\tilde{X}$ and $\tilde{N}$ are independent random variables, the mutual information $I(\tilde{X}; \tilde{Y})$ equals to

$$I(\tilde{X}; \tilde{Y}) = \frac{1}{2}\log\left(\frac{Var(\tilde{Y})}{Var(\tilde{N})}\right). \tag{2.121}$$

Since $\tilde{X}$ and $\tilde{N}$ are independent random variables, we can write that

$$Var(\tilde{Y}) = Var(\tilde{X}) + Var(\tilde{N}). \tag{2.122}$$

Substituting (2.122) into (2.121), we get

$$I(\tilde{X}; \tilde{Y}) = \frac{1}{2}\log\left(1 + \frac{Var(\tilde{X})}{Var(\tilde{N})}\right). \tag{2.123}$$

If we choose $\tilde{N}$ as $\tilde{N} \sim N\left(0, \sigma^2_{\tilde{N}}\right)$, then $Var(\tilde{N}) = E(\tilde{N}^2)$. This means that $Var(\tilde{N})$ is nothing but the average noise power, i.e., $S_N = E(\tilde{N}^2)$. If we choose the Gaussian random variable $\tilde{X}$ with zero mean, i.e., $\tilde{X} \sim N\left(0, \sigma^2_{\tilde{X}}\right)$ then $Var(\tilde{X}) = E(\tilde{X}^2)$ which is nothing but the average power of $\tilde{X}$, and let's denote this power by $S_X = E(\tilde{X}^2)$. In this case, (2.123) is modified as

$$I(\tilde{X}; \tilde{Y}) = \frac{1}{2}\log\left(1 + \frac{S_X}{S_N}\right) \tag{2.124}$$

where it is seen that as S_X increases so do $I(\tilde{X}; \tilde{Y})$. However, in practical applications we are usually interested in signals having constant powers. For this reason, we should consider the maximum value of the mutual information under power constraint.

Definition The Gaussian channel is described in Fig. 2.33 where $\tilde{X}$ is the input of the channel, $\tilde{Y}$ is the output of the channel and $\tilde{N}$ is the noise with zero mean Gaussian distribution. The relation between input and output is described by $\tilde{Y} = \tilde{X} + \tilde{N}$.

Fig. 2.33 Gaussian channel

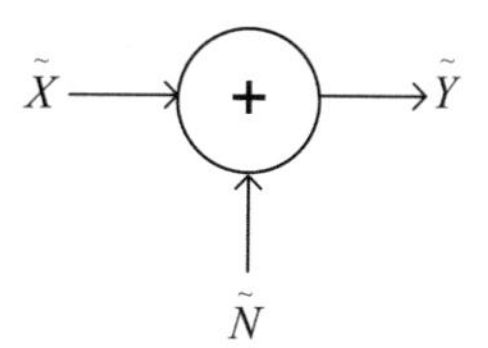

Definition The capacity of the Gaussian channel with input power constraint P is defined as

$$C = \max_{\substack{f(x) \\ E(\tilde{X}^2) \leq P}} I(\tilde{X}; \tilde{Y}) \tag{2.125}$$

where $f(x)$ is the distribution of the input random variable $\tilde{X}$. In our previous examples, we found that the mutual information in (2.125) gets its maximum value if $\tilde{X}$ is Gaussian distributed. And if $\tilde{X}$ has zero mean, in this case, the maximum value of the mutual information is expressed as in (2.124), i.e.,

$$C = \frac{1}{2}\log\left(1 + \frac{S_X}{S_N}\right) \text{ bits/tranmission.} \tag{2.126}$$

The mutual information in (2.126) is for the transmission of one sample of the continuous random variable, i.e., for the transmission of one sample of the continuous signal $\tilde{X}$.

In practical applications, a continuous signal time bandlimited to B Hz is sampled by the sampling frequency $f_s = 2B$ Hz, this means that, every per-second, $2B$ samples are taken from continuous time signal, and these samples are transmitted. In this case, the capacity expression in (2.126) can be expressed in terms of per-second unit as

$$C = 2B \times \frac{1}{2}\log\left(1 + \frac{S_X}{S_N}\right) \text{ bits/s} \rightarrow C = B\log\left(1 + \frac{S_X}{S_N}\right) \text{ bits/s} \tag{2.127}$$

which is also known as Shannon capacity for AWGN channels.

Example 2.41 The bandwidth of a lowpass continuous time signal is $B = 100$ Hz. The average power of this signal is $S_X(\text{dB}) = 20$ dB. The continuous time signal is sampled and the samples are transmitted through an AWGN channel whose average noise power is $S_N(\text{dB}) = 0$ dB. Calculate the channel capacity in terms of bits/tranmission and bits/s.

Solution 2.41 The average power of the signal can be calculated as

$$20 = 10\log_{10} S_X \rightarrow S_X = 100 \text{ W}.$$

In a similar manner the average noise power can be calculated as

$$0 = 10\log_{10} S_N \rightarrow S_N = 1 \text{ W}.$$

Then, the AWGN channel capacity can be calculated as in

$$C = \frac{1}{2}\log\left(1 + \frac{S_X}{S_N}\right) \rightarrow C = \frac{1}{2}\log\left(1 + \frac{100}{1}\right) \rightarrow C = 3.33 \text{ bits/transmission}$$

or

$$C = 3.33 \text{ bits/sample}.$$

The continuous time lowpass signal has the bandwidth $B = 100$ Hz. When this signal is samples, we take per second at least $f_s = 2 \times 100 = 200$ samples from this signal and transmit them through AWGN channel. Thus, for the transmission of 200 sampled per second, the channel capacity happens to be

$$C = 200 \times \frac{1}{2}\log\left(1 + \frac{100}{1}\right) \rightarrow C = 666 \text{ bits/s}.$$

Example 2.42 For the previous example, is it possible to increase the capacity value without increasing power or bandwidth of the signal?

Solution 2.42 The capacity per-sample transmission is

$$C = \frac{1}{2}\log\left(1 + \frac{S_X}{S_N}\right).$$

The sampling frequency is chosen as $f_s \geq 2B$ which indicates the number of samples transmitted per-second. In our previous example, we can decide on the sampling frequency considering the criteria

$$f_s \geq 2 \times 100 \rightarrow f_s \geq 200.$$

If we choose the sampling frequency as

$$f_s = 400 \text{ Hz}$$

the transmission speed per-second happens to be as

$$C = 400 \times \frac{1}{2}\log\left(1 + \frac{S_X}{S_N}\right) \rightarrow C = 400 \times 3.33 \rightarrow C = 1332 \text{ bits/s}.$$

2.5 Bounds and Limiting Cases on AWGN Channel Capacity

In this part, we will inspect the effect of bandwidth of the information signal, and signal-to-noise ratio on the AWGN channel capacity.

2.5.1 Effect of Information Signal Bandwidth on AWGN Channel Capacity

If the AWGN channel capacity expression in (2.127) is inspected in details, it is seen that as the bandwidth of the continuous time signal increases, the capacity also increases, since the bandwidth is a multiplicative term in the capacity expression. However, the actual scenario is not like that. This is due to the noise power. As the bandwidth of the signal increases, the noise power increases as well. This is due to the dependency of the noise power to the bandwidth of the signal.

The white-noise's double sided power spectral density can be expressed as

$$S_d(f) = \frac{N_0}{2}.$$

The average noise power appearing in AWGN channel in this case can be expressed as

$$S_N = 2B \times S_d(f) \rightarrow S_N = 2B \times \frac{N_0}{2} \rightarrow S_N = BN_0. \tag{2.128}$$

If (2.128) is substituted into (2.127), we obtain

$$C = B\log\left(1 + \frac{S_X}{BN_0}\right)$$

which can be written as

$$C = \frac{\log\left(1 + \frac{S_X}{N_0}\frac{1}{B}\right)}{\frac{1}{B}}. \tag{2.129}$$

Now, let's consider the limiting case of (2.129) as B goes to infinity. In this case, we have

$$C_\infty = \lim_{B\to\infty} \frac{\log\left(1+\frac{S_X}{N_0}\frac{1}{B}\right)}{\frac{1}{B}} \to C_\infty = \frac{0}{0}$$

which is the 0/0 uncertainty. We can avoid this uncertainty by taking the derivative of numerator and denominator with respect to B. If we take the derivative of numerator and denominator with respect to B, we get

$$C_\infty = \lim_{B\to\infty} \frac{\frac{\partial \log\left(1+\frac{S_X}{N_0}\frac{1}{B}\right)}{\partial B}}{\frac{\partial \frac{1}{B}}{\partial B}} \to C_\infty = \lim_{B\to\infty} \frac{\frac{\left(-\frac{S_X}{N_0}\frac{1}{B^2}\right)}{\left(1+\frac{S_X}{N_0}\frac{1}{B}\right)}\left(\frac{1}{\ln 2}\right)}{-\frac{1}{B^2}} \to$$
$$C_\infty = \lim_{B\to\infty} \frac{\left(\frac{S_X}{N_0}\right)}{\left(1+\frac{S_X}{N_0}\frac{1}{B}\right)}\left(\frac{1}{\ln 2}\right) \tag{2.130}$$

which yields

$$C_\infty = \frac{S_X}{N_0}\log e \to C_\infty = 1.44\frac{S_X}{N_0}. \tag{2.131}$$

When the result of the L' Hospitals' rule in (2.130) is evaluated for $B = 0$, we get

$$C_0 = \lim_{B\to 0} \frac{\left(\frac{S_X}{N_0}\right)}{\left(1+\frac{S_X}{N_0}\frac{1}{B}\right)}\left(\frac{1}{\ln 2}\right) \to C_0 = 0. \tag{2.132}$$

Thus, we can conclude that as the bandwidth goes to zero, so does the capacity, and as the bandwidth goes to infinity, capacity saturates at a constant value. The effect of signal bandwidth on AWGN channel capacity is illustrated in Fig. 2.34.

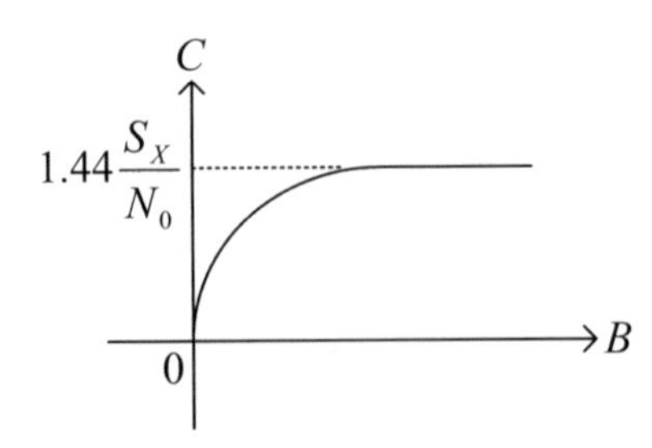

Fig. 2.34 Effect of bandwidth on channel capacity

Note:

$$\frac{\partial \log f(x)}{\partial x} = \frac{f'(x)}{f(x)} \frac{1}{\ln a}$$

For AWGN channels, the capacity is usually expressed in terms of bits/s unit. This unit is related to the number of samples generated per-second from the continuous time signal. Once, we generate data samples, we add digital noise signal to these data samples. The maximum value of the mutual information or channel capacity is defined between noisy samples and original samples generated for one-second time duration.

In communication engineering, capacity is accepted as the maximum value of the communication speed, i.e., the maximum value of the mutual information between input and output of the communication channel per-second.

2.5.2 *Effect of Signal to Noise Ratio on the Capacity of AWGN Channel*

Before studying effect of signal to noise ratio on the capacity of AWGN channel, let's solve some calculus problems to prepare ourselves for the subject.

Example 2.43 If

$$f(r) = \frac{2^r - 1}{r},$$

calculate the limit

$$\lim_{r \to 0} f(r).$$

Solution 2.43 If the limit is evaluated for $r = 0$, we get

$$\lim_{r \to 0} f(r) = \frac{0}{0}$$

where 0/0 uncertainty occurs. The uncertainty can be avoided by taking the derivative of numerator and denominator and evaluating the limit as in

$$\lim_{r \to 0} f(r) = \lim_{r \to 0} \frac{\partial f(r)}{\partial r} \rightarrow \lim_{r \to 0} f(r) = \lim_{r \to 0} \frac{2^r \ln 2}{1} \rightarrow \lim_{r \to 0} f(r) = \ln 2$$

where $\ln 2 = 0.693$ which can be expressed in dB as

$$\ln 2 = 0.693 \rightarrow 10\log_{10} 0.693 = -1.6 \text{ dB}$$

In summary,

$$\lim_{r\rightarrow 0} f(r) = \ln 2 = 0.693 = -1.6 \text{ dB}. \tag{2.133}$$

Example 2.44 Draw the graph of

$$f(r) = \frac{2^r - 1}{r}$$

for $r \geq 0$, and determine the regions $y > f(r)$, and $y < f(r)$.

Solution 2.44 Giving values to r the graph of $f(r)$ can be drawn as in Fig. 2.35. To determine the regions $y > f(r)$ and $y < f(r)$ we can consider a value of r, i.e., r_0 and determine the value of $f(r_0)$ and consider those values of y greater than or smaller than $f(r_0)$, i.e., those values of y satisfying $y > f(r_0)$ and those values of y satisfying $y < f(r_0)$. The regions for $y > f(r)$ and $y < f(r)$ are depicted in Figs. 2.36 and 2.37.

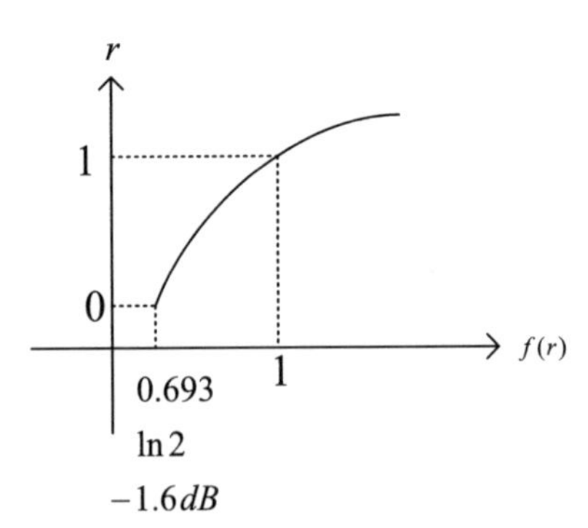

Fig. 2.35 The graph of $f(r)$

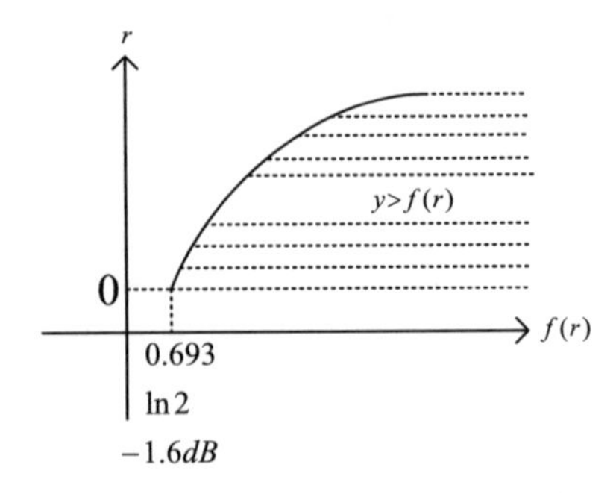

Fig. 2.36 $y > f(r)$ region

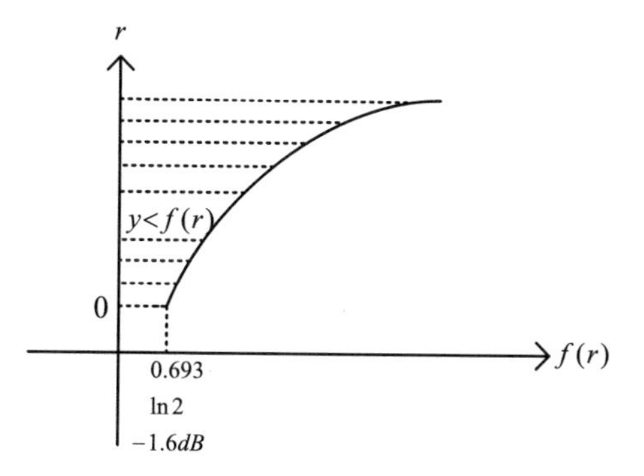

Fig. 2.37 $y<f(r)$ region

Now, we can return to our subject. Let R be the transmission speed for an AWGN channel per second, i.e., the value of the mutual information per second. In any practical communication system, we must have $R<C$. For the AWGN channel, $R<C$ implies that

$$R<C \rightarrow R<B\log\left(1+\frac{S_X}{BN_0}\right)$$

which can be written as

$$\frac{R}{B}<\log\left(1+\frac{S_X}{BN_0}\right). \tag{2.134}$$

Let $r=R/B$ and E_b be the energy per-bit. The signal energy in (2.134) can be expressed as $S_X=RE_b$. With the given definitions, (2.134) can be written as

$$\underbrace{\frac{R}{B}}_{=r}<\log\left(1+\underbrace{\frac{R}{B}}_{=r}\frac{E_b}{N_0}\right) \tag{2.135}$$

which is re-written using parameter r as in

$$r<\log\left(1+r\frac{E_b}{N_0}\right). \tag{2.136}$$

From (2.136), we get

$$\frac{E_b}{N_0}>\frac{2^r-1}{r} \tag{2.137}$$

which is a lower bound for E_B/N_0 obtained starting from $R < C$. If we take the limit of (2.137) as r goes to 0, we obtain

$$\frac{E_b}{N_0} > 0.693 = \ln 2 = -1.6 \text{ dB}$$

which can be interpreted as the minimum required signal-to-noise ratio for reliable communication, i.e., for reliable communication, we must have

$$\frac{E_b}{N_0} > -1.6 \text{ dB}. \tag{2.138}$$

In Figs. 2.38 and 2.39, for a given rate r, the reliable and unreliable communication regions are illustrated considering E_b/N_0 values.

Example 2.45 For a given communication system if

$$\frac{E_b}{N_0} = 0.5$$

is it possible to do a reliable communication?

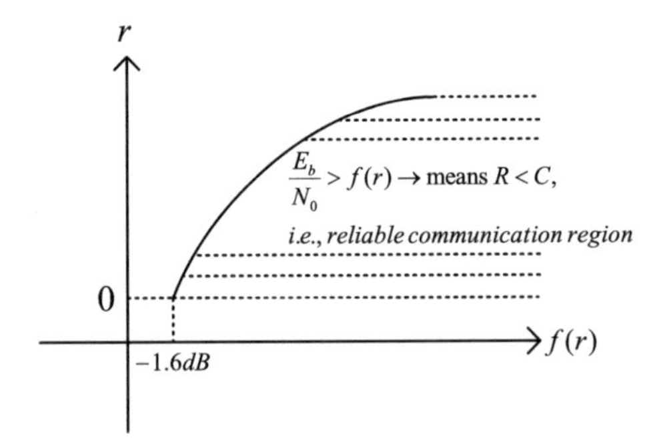

Fig. 2.38 Reliable communication region

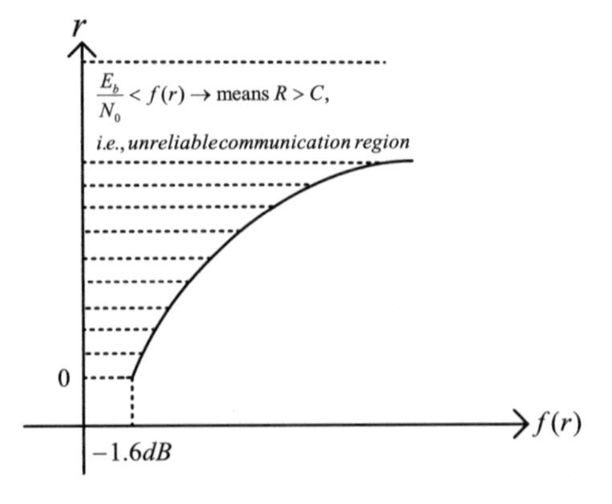

Fig. 2.39 Unreliable communication region

Solution 2.45 For a reliable communication we must have

$$\frac{E_b}{N_0} \geq 0.693.$$

Since, $\frac{E_b}{N_0} = 0.5$ it is not possible to do a reliable communication.

Example 2.46 We want to transmit 8000 bits/s through an AWGN channel. The sampled analog signal has the bandwidth $B = 1000$ Hz. What should be the minimum E_b/N_0 value for reliable transmission?

Solution 2.46 For reliable transmission, we should have

$$\frac{E_b}{N_0} > \frac{2^r - 1}{r} \tag{2.139}$$

where

$$r = \frac{R}{B}$$

which can be evaluated for the given values in the example as

$$r = \frac{8000}{1000} \rightarrow r = 8.$$

Then, the bound in (2.139) can be computed as

$$\frac{E_b}{N_0} > \frac{2^8 - 1}{8} \rightarrow \frac{E_b}{N_0} > \frac{2^8 - 1}{8} \rightarrow \frac{E_b}{N_0} > 32 \rightarrow \frac{E_b}{N_0} > 15 \text{ dB}.$$

Thus, we should have

$$\frac{E_b}{N_0} > 15 \text{ dB}$$

to make a reliable transmission at the speed 8000 bits/s for the given signal bandwidth $B = 1000$ Hz.

Exercise For the communication through AWGN channel, if the sampling frequency for the continuous time signal is chosen as $f_s = 4B$ where B is the bandwidth of the lowpass signal, find the minimum required E_b/N_0 for reliable communication.

Hint: The capacity of the AWGN channel in this case equals to

$$C = 4B \times \frac{1}{2}\log\left(1 + \frac{S_X}{\frac{N_0}{2} \times 4B}\right) \rightarrow C = 2B\log\left(1 + \frac{S_X}{N_0 2B}\right)$$

and for reliable communication, we must have $R < C$ where R is the transmission speed, i.e., value of the mutual information per-second.

Problems

(1) Calculate the entropy of the Rayleigh random variable whose probability density function is given as

$$f(x) = \frac{x}{\sigma^2}\exp\left(-\frac{x^2}{2\sigma^2}\right).$$

(2) The random variables $\tilde{X}_1$ and $\tilde{X}_2$ have Gaussian distributions such that

$$\tilde{X}_1 \sim N\left(0, \sigma^2\right) \quad \tilde{X}_2 \sim N\left(\mu, \sigma^2\right).$$

Calculate the probabilistic distance between $\tilde{X}_1$ and $\tilde{X}_2$, i.e., $D(\tilde{X}_1||\tilde{X}_2) = ?$

(3) Find the joint entropy of two exponentially distributed continuous random variable $\tilde{X}_1$ and $\tilde{X}_2$ whose marginal distributions are given as

$$f(x) = \begin{cases} \lambda_1 e^{-\lambda_1 x} & x \geq 0 \\ 0 & otherwise \end{cases} \quad g(x) = \begin{cases} \lambda_2 e^{-\lambda_2 x} & x \geq 0 \\ 0 & otherwise. \end{cases}$$

(4) Find the differential entropy of the continuous random variable $\tilde{X}$ whose probability density function is given as

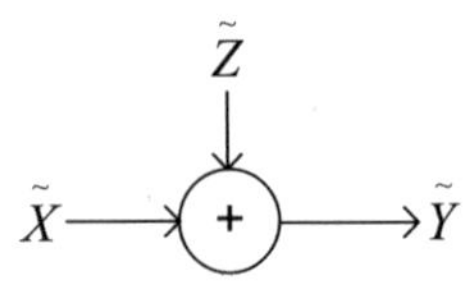

Fig. 2.P1 Additive communication channel

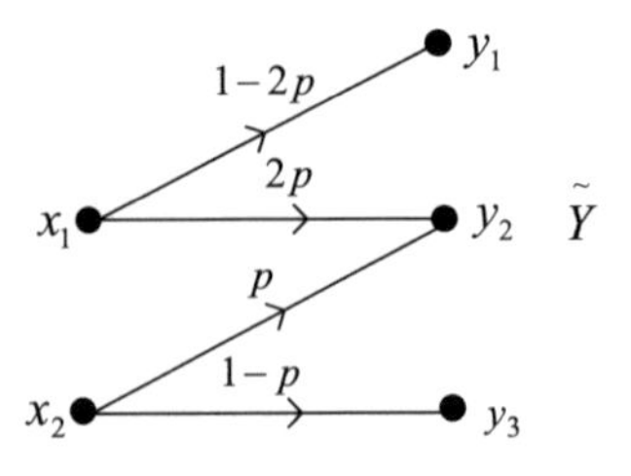

Fig. 2.P2 Discrete memoryless channel

$$f(x) = \begin{cases} 2x & 0 \leq x \leq 1 \\ 0 & otherwise. \end{cases}$$

(5) For the additive communication channel shown in Fig. 2.P1, the random variable $\tilde{Z}$ is exponentially distributed, $\tilde{X}$ and $\tilde{Z}$ are independent of each other. Find the capacity of this channel.

(6) For the discrete memoryless channel shown in Fig. 2.P2, the transition probability p is a constant number. Calculate the channel capacity of the channel in Fig. 2.P2.

Chapter 3
Typical Sequences and Data Compression

In this chapter, we will first provide information about typical sequences the introduce the data compression subject. In fact, the data compression subject is a direct consequence of the typical sequences. For this reason, it is very important to comprehend the concept of typical sequences to understand the data compression phenomenon.

3.1 Independent Identically Distributed Random Variables (IID Random Variables)

Assume that we perform an experiment repeatedly. At each perform of the experiment we get some results, and these results come from the same sample space. This means that the trial of each experiment can be considered as the occurrence of a random variable, and these random variables have the same sample space, but, they are independent of each other. This type of phenomenon gives rise to the concept of *independent and identically distributed* (*i.i.d. or IID*) random variables.

Example 3.1 Assume that we have an unbiased coin, and we flip the coin five times. In this case, we consider that we have five random variables independent of each other, and these random variables have the same sample space, and have the same probability mass function. That it, we have the random variables

$$\tilde{X}_1, \tilde{X}_2, \tilde{X}_3, \tilde{X}_4, \tilde{X}_5,$$

and these random variables have the same range set, i.e.,

O. Gazi, *Information Theory for Electrical Engineers*, Signals and Communication Technology, https://doi.org/10.1007/978-981-10-8432-4_3

$$R_{\tilde{X}_1} = \{h, t\}, \quad R_{\tilde{X}_2} = \{h, t\}, \quad R_{\tilde{X}_3} = \{h, t\}, \quad R_{\tilde{X}_4} = \{h, t\}, \quad R_{\tilde{X}_5} = \{h, t\}$$

and these random variables have the same probability mass functions. Let's denote the probability mass functions of theses random variables as

$$p_1(x), \quad p_2(x), \quad p_3(x), \quad p_4(x), \quad p_5(x)$$

and we have

$$p_1(h) = \frac{1}{2} \quad p_1(t) = \frac{1}{2} \quad p_2(h) = \frac{1}{2} \quad p_2(t) = \frac{1}{2} \quad p_3(h) = \frac{1}{2} \quad p_3(t) = \frac{1}{2}$$
$$p_4(h) = \frac{1}{2} \quad p_4(t) = \frac{1}{2} \quad p_5(h) = \frac{1}{2} \quad p_5(t) = \frac{1}{2}$$

Since these random variables have the same probability mass functions, we can denote the common probability mass function by the generic notation $p(x)$. To indicate that these random variables have the common distribution $p(x)$, we use the expression

$$\tilde{X}_1, \tilde{X}_2, \tilde{X}_3, \tilde{X}_4, \tilde{X}_5 \sim p(x). \tag{3.1}$$

In addition, we can indicate a generic random variable with the same probability mass function as in

$$\tilde{X} \sim p(x). \tag{3.2}$$

Example 3.2 Let's define three IID random variables $\tilde{X}_1, \tilde{X}_2, \tilde{X}_3$ and a generic random variable $\tilde{X}$ all having the same probability mass function and the same range set as in

$$\tilde{X} \sim p(x), \quad R_{\tilde{X}} = \{a, b, c\}, \quad p(a) = \frac{1}{4}, \quad p(b) = \frac{2}{4}, \quad p(c) = \frac{1}{4}$$
$$\tilde{X}_1 \sim p(x), \quad R_{\tilde{X}_1} = \{a, b, c\}, \quad p_1(a) = \frac{1}{4}, \quad p_1(b) = \frac{2}{4}, \quad p_1(c) = \frac{1}{4}$$
$$\tilde{X}_2 \sim p(x), \quad R_{\tilde{X}_2} = \{a, b, c\}, \quad p_2(a) = \frac{1}{4}, \quad p_2(b) = \frac{2}{4}, \quad p_2(c) = \frac{1}{4}$$
$$\tilde{X}_3 \sim p(x), \quad R_{\tilde{X}_3} = \{a, b, c\}, \quad p_3(a) = \frac{1}{4}, \quad p_3(b) = \frac{2}{4}, \quad p_3(c) = \frac{1}{4}$$

and we indicate the IID random variables as

$$\tilde{X}_1, \tilde{X}_2, \tilde{X}_3 \sim p(x).$$

Definition The random variables $\tilde{X}_1, \tilde{X}_2, \ldots, \tilde{X}_N$ are IID if

$$F_{\tilde{X}_i}(x) = F_{\tilde{X}}(x) \quad \forall i = 1, 2, \ldots, N$$

where $F_{\tilde{X}_i}(x)$ is the cumulative distribution function of $\tilde{X}_i$, and $F_{\tilde{X}}(x)$ is the cumulative distribution function of $\tilde{X}$ and

$$F_{\tilde{X}_1, \tilde{X}_2, \ldots, \tilde{X}_N}(x) = \prod_{i=1}^{N} F_{\tilde{X}_i}(x_i) \tag{3.3}$$

where $F_{\tilde{X}_1, \tilde{X}_2, \ldots, \tilde{X}_N}(x)$ is the joint cumulative distribution function of the random variables $\tilde{X}_1, \tilde{X}_2, \ldots, \tilde{X}_N$.

3.1.1 *The Weak Law of Large Numbers*

Let the random variables $\tilde{X}_1, \tilde{X}_2, \ldots, \tilde{X}_N$ be IID, and

$$\tilde{X}_1, \tilde{X}_2, \ldots, \tilde{X}_N \sim p(x).$$

The weak law of large numbers indicates that for IID random variables $\tilde{X}_1, \tilde{X}_2, \ldots, \tilde{X}_N$, we have

$$\frac{1}{N} \sum_{i=1}^{N} x_i \approx E(\tilde{X}) \tag{3.4}$$

where x_i is the value produced by the random variable $\tilde{X}_i$ and $E(\tilde{X})$ is the probabilistic average of the generic random variable having probability mass function $p(x)$.

Example 3.3 The discrete random variable $\tilde{X}$ has the probability mass function $p(x)$ defined as

$$p(x = -1) = 0.2 \quad p(x = 1) = 0.8.$$

The IID random variables $\tilde{X}_1, \tilde{X}_2, \ldots, \tilde{X}_N$ have probability mass function $p(x)$, i.e.,

$$\tilde{X}_1, \tilde{X}_2, \ldots, \tilde{X}_N \sim p(x).$$

The expected value of $\tilde{X}$ is calculated as

$$E(\tilde{X}) = \sum_{x} xp(x) \rightarrow E(\tilde{X}) = -1 \times 0.2 + 1 \times 0.8 \rightarrow E(\tilde{X}) = 0.6.$$

According to the weak law of large numbers, we have

$$\frac{1}{N}\sum_{i=1}^{N} x_i \approx E(\tilde{X}) \rightarrow \frac{1}{N}\sum_{i=1}^{N} x_i \approx 0.6.$$

3.2 Convergence of Random Variable Sequences

Before studying the convergence of random numbers, let's first remember the convergence of real numbers. If we have a sequence of real numbers, we wonder whether the sequence converges or not. For example, the sequence

$$\frac{2}{3}, \frac{4}{4}, \frac{6}{5}, \ldots$$

can be expressed as

$$x_N = \frac{2N}{N+2}, \quad \text{for } N = 1,\ 2,\ \cdots$$

which converges to 2. We say that the sequence $x_1,\ x_2, \ldots$ denoted by x_N converges to L if

$$\lim_{n \rightarrow \infty} x_N = L \tag{3.5}$$

which means that there exists $\epsilon > 0$, and $N \in \mathrm{N}$ such that

$$|x_N - L| < \epsilon \quad \text{for } N > N_0. \tag{3.6}$$

Now, let's consider the convergence of random variables. A random variable is nothing but a real valued function defined on the simple outcomes of an experiment. Let's denote sample space of an experiment consisting of simple outcomes by

$$S = \{s_1,\ s_2, \ldots s_M\}$$

where s_i's are nothing but the simple outcomes of an experiment. A random variable $\tilde{X}$ is a real valued function defined on the simple outcomes as in

$$\tilde{X}(s_i) = x_i, \quad i = 1, \ldots, M.$$

Example 3.4 Consider the coin toss experiment. The sample space of this experiment is $S = \{H, T\}$. A sequence of random variables $\tilde{X}_1, \tilde{X}_2, \ldots$ on this sample space can be defined as

$$\tilde{X}_N(s) = \begin{cases} \frac{N}{2N+1} & \text{if } s = H \\ \frac{N}{N+3} & \text{if } s = T. \end{cases}$$

We see that as N goes to infinity the random variable sequence converges to the random variable

$$\tilde{X}(s) = \begin{cases} \frac{1}{2} & \text{if } s = H \\ 1 & \text{if } s = T. \end{cases}$$

3.2.1 *Different Types of Convergence for the Sequence of Random Variables*

Before studying the convergence of sequence of random variables, let's first remember what a random variable was.

An experiment is a physical event, and at every trial of the experiment we get some outputs. The outputs of the experiments are called simple outcomes. Thc set consisting of distinct simple outcomes is called the sample space which is denoted by S. The subsets of the sample space are called events. If we consider two experiments with sample spaces S_1 and S_2 together, then the combined experiment has the sample space $S_1 \times S_2$ where $\times$ indicates the Cartesian product of two sample spaces.

A random variable $\tilde{X}(\cdot)$ is a real valued function, its inputs are the simple outcomes, and its outputs are real numbers. The random variable $\tilde{X}(\cdot)$ can also be considered as a mapping function among distinct simple outcomes, i.e., elements of the sample space, and real numbers.

Let's give some simple examples to remind the random variable concept.

Example 3.5 Experiment-1: Toss of a fair coin. *Simple Outcomes*: H, T. *Sample Space*: $S_1 = \{H, T\}$.

Experiment-2: Toss of a fair die. *Simple Outcomes*: $F_1, F_2, F_3, F_4, F_5, F_6$. *Sample Space* $S_2 = \{F_1, F_2, F_3, F_4, F_5, F_6\}$.

Combined experiment: Toss of a fair coin and a fair die. *Sample Space*:

$$S = \{HF_1,\ HF_2,\ HF_3, HF_4,\ HF_5,\ HF_6,\ TF_1,\ TF_2,\ TF_3, TF_4,\ TF_5,\ TF_6\}$$

Random Variable $\tilde{X}(\cdot)$ on S is defined as

$$\tilde{X}(s) = \begin{cases} 1 & \textit{if } s \text{ includes } H \text{ and an even indexed face} \\ 2 & \textit{if } s \text{ includes } T \text{ and an odd indexed face} \\ -1 & \text{otherwise.} \end{cases}$$

Then, we have

$$\begin{aligned} &\tilde{X}(HF_1) = -1 \quad \tilde{X}(HF_2) = 1 \quad \tilde{X}(HF_3) = -1 \quad \tilde{X}(HF_4) = 1 \\ &\tilde{X}(HF_5) = -1 \quad \tilde{X}(HF_6) = 1 \\ &\tilde{X}(TF_1) = 2 \quad \tilde{X}(TF_2) = -1 \quad \tilde{X}(TF_3) = 2 \quad \tilde{X}(TF_4) = -1 \\ &\tilde{X}(TF_5) = 2 \quad \tilde{X}(TF_6) = -1. \end{aligned}$$

The range set of the random variable $\tilde{X}$ can be written as

$$R_{\tilde{X}} = \{-1,\ 1,\ 2\}.$$

The probability mass function $p(x)$ for the random variable function $\tilde{X}$ is defined as

$$p(x) = Prob\{s|\tilde{X}(s) = x\}$$

where $\{s|\tilde{X}(s) = x\}$ indicates a subset of the sample space S, i.e., the subset consisting of the simple outcomes s which satisfy $\tilde{X}(s) = x$.

Convergence for the Sequence of Random Variables:

There are four types of convergence of sequence of random variables. These types are

(a) Convergence in Distribution
(b) Convergence in Probability
(c) Convergence in Mean Square Sense
(d) Almost Surely Convergence.

Let's briefly explain these convergence types.

Convergence in Distribution:

A sequence of random variables $\tilde{X}_1, \tilde{X}_2, \ldots, \tilde{X}_N$, converges in distribution to a random variable $\tilde{X}$, if we have

$$\lim_{N\to\infty} F_{\tilde{X}_N}(x) = F_{\tilde{X}}(x) \tag{3.7}$$

and the convergence in distribution is indicated by

$$\tilde{X}_N \xrightarrow{d} \tilde{X}. \tag{3.8}$$

Example 3.6 For the sequence of random variables $\tilde{X}_1, \tilde{X}_2, \ldots, \tilde{X}_N$, the cumulative distribution function is given as,

$$F_{\tilde{X}_N}(x) = 1 - \left(1 - \frac{\lambda x}{N}\right)^N.$$

Then, as $N \to \infty$ we get

$$\lim_{N\to\infty} F_{\tilde{X}_N}(x) = 1 - e^{-\lambda x}.$$

Thus, we have

$$\tilde{X}_N \xrightarrow{d} \tilde{X}, \quad \text{where } \tilde{X} \sim f(x) = \lambda e^{-\lambda x}.$$

Convergence in Probability:

A sequence of random variables $\tilde{X}_1, \tilde{X}_2, \ldots, \tilde{X}_N$ converges to $\tilde{X}$ in probability if for any $\epsilon > 0$

$$\lim_{N\to\infty} \text{Prob}\left(\left|\tilde{X}_N - \tilde{X}\right| > \epsilon\right) = 0 \tag{3.9}$$

and the convergence in distribution is indicated by

$$\tilde{X}_N \xrightarrow{p} \tilde{X}. \tag{3.10}$$

Example 3.7 Considering sequence of random variables $\tilde{X}_1, \tilde{X}_2, \ldots, \tilde{X}_N$, the random variable $\tilde{X}_N$ is defined as

$$\tilde{X}_N = \left(1 + \frac{1}{N}\right)\tilde{X} \tag{3.11}$$

where the random variable $\tilde{X}$ has the range set $R_{\tilde{X}} = \{-1, 1\}$ with the probability mass function defined as

$$p(x = -1) = \frac{1}{3} \quad p(x = 1) = \frac{2}{3}.$$

From (3.11), we get

$$|\tilde{X}_N - \tilde{X}| = \left|\frac{\tilde{X}}{N}\right|$$

where $\tilde{X} = 1$ or $\tilde{X} = -1$, i.e., $\tilde{X} = 1$. Then, it is clear that

$$\lim_{N\to\infty} Prob\left(|\tilde{X}_N - \tilde{X}| = \frac{1}{N} > \epsilon\right) = 0.$$

Thus, we have

$$\tilde{X}_N \xrightarrow{p} \tilde{X}.$$

Convergence in Mean Square Sense:

A sequence of random variables $\tilde{X}_1, \tilde{X}_2, \ldots, \tilde{X}_N$ converges to $\tilde{X}$ in mean square sense if

$$\lim_{N\to\infty} \mathrm{E}\left(|\tilde{X}_N - \tilde{X}|^2\right) = 0. \tag{3.12}$$

and the convergence in mean square sense is indicated by

$$\tilde{X}_N \xrightarrow{m.s.} \tilde{X}. \tag{3.13}$$

Example 3.8 Consider a sequence of IID random variables $\tilde{X}_1, \tilde{X}_2, \ldots, \tilde{X}_N$ each having mean value μ and variance σ^2. Consider the random variable sequence $\tilde{Y}_N$ defined as

$$\tilde{Y}_N = \frac{1}{N}\sum_{i=1}^{N}\tilde{X}_i$$

and the constant random variable $\tilde{X}$ with range set $R_{\tilde{X}} = \{\mu\}$. Let's show that

$$\lim_{N\to\infty} \mathrm{E}\left(\left|\tilde{Y}_N - \tilde{X}\right|^2\right) = 0.$$

The mean of the random variable $\tilde{Y}_N$ can be calculated as

$$E(\tilde{Y}_N) = \frac{1}{N}\sum_{i=1}^{N}E(\tilde{X}_i) \to E(\tilde{Y}_N) = \frac{N\mu}{N} \to E(\tilde{Y}_N) = \mu.$$

Since $\tilde{X}$ is a constant random variable, the difference $\left|\tilde{Y}_N - \tilde{X}\right|^2$ can be written as

$$\left|\tilde{Y}_N - \mu\right|^2.$$

Then,

$$\mathrm{E}\left(\left|\tilde{Y}_N - \tilde{X}\right|^2\right) = \mathrm{E}\left(\left|\tilde{Y}_N - \mu\right|^2\right) \to \mathrm{E}\left(\left|\tilde{Y}_N - \tilde{X}\right|^2\right) = Var(\tilde{Y}_N) \tag{3.14}$$

where $Var(\tilde{Y}_N)$ can also be written as

$$Var(\tilde{Y}_N) = Var\left(\frac{1}{N}\sum_{i=1}^{N}\tilde{X}_i\right)$$

leading to

$$Var(\tilde{Y}_N) = \frac{1}{N^2}\sum_{i=1}^{N}Var(\tilde{X}_i)$$

where using $Var(\tilde{X}_i) = \sigma^2$, we can write (3.14) as

$$\mathrm{E}\left(\left|\tilde{Y}_N - \tilde{X}\right|^2\right) = \frac{1}{N^2}N\sigma^2 \to \mathrm{E}\left(\left|\tilde{Y}_N - \tilde{X}\right|^2\right) = \frac{\sigma^2}{N}$$

whose limit as $N \to \infty$ equals to

$$\lim_{N\to\infty} \mathrm{E}\left(\left|\tilde{Y}_N - \tilde{X}\right|^2\right) = 0$$

which indicates that $\tilde{Y}_N$ converges to $\tilde{X} = \mu$ in mean square sense.

Almost Surely Convergence or Convergence with Probability 1:

The random variable sequence $\tilde{X}_N$ converges to $\tilde{X}$ almost surely if

$$\mathrm{Prob}\left(s \mid \lim_{N\to\infty} \tilde{X}_N(s) = \tilde{X}(s)\right) = 1 \tag{3.15}$$

where s is a simple outcome of the sample space, and almost surely convergence is indicated by

$$\tilde{X}_N \overset{a.s.}{\to} \tilde{X}. \tag{3.16}$$

Example 3.9 The sample space is given as $S = [0\,1]$. The probability measure of the event $E = [a\,b]$, $a < b$ on this sample space is given as

$$Prob(E) = b - a.$$

The random variable sequence $\tilde{X}_N$ on the sample space is defined as

$$\tilde{X}_N(s) = \begin{cases} 2 & 0 \le s \le \frac{N+1}{2N} \\ 0 & otherwise \end{cases}$$

where s is an interval included in sample space S, i.e., s is an event. The random variable $\tilde{X}$ on this sample space is defined as

$$\tilde{X}(s) = \begin{cases} 2 & 0 \le s < \frac{1}{2} \\ 0 & otherwise. \end{cases}$$

Let's show that

$$\tilde{X}_N(s) \overset{a.s.}{\to} \tilde{X}.$$

Let $B = \left(s \mid \lim_{N\to\infty} \tilde{X}_N(s) = \tilde{X}(s)\right)$. B indicates the intervals s such that $\lim_{N\to\infty} \tilde{X}_N(s) = \tilde{X}(s)$.

For $s < \frac{1}{2}$, i.e., $s \in \left[0\ \frac{1}{2}\right)$ we have

$$\tilde{X}_N(s) = \tilde{X}(s) = 1.$$

And when $s > \frac{1}{2}$, we have

$$\lim_{N \to \infty} \tilde{X}_N(s) = \tilde{X}(s) = 0.$$

Then, the event B can be written as

$$B = \left[0\ \frac{1}{2}\right) \cup \left(\frac{1}{2}\ 1\right] \to B = S - \frac{1}{2}$$

Since $Prob(B) = 1$, then we have

$$\tilde{X}_N(s) \xrightarrow{a.s.} \tilde{X}.$$

3.3 Asymptotic Equipartition Property Theorem

Let IID random variables have the probability mass function $p(x)$, i.e.,

$$\tilde{X}_1, \tilde{X}_2, \ldots, \tilde{X}_N \sim p(x).$$

The joint probability mass function of these IID random variables is indicated by

$$p(x_1, x_2, \ldots, x_N)$$

i.e.,

$$p(x_1, x_2, \ldots, x_N) = Prob\left(\tilde{X}_1 = x_1, \tilde{X}_2 = x_2, \ldots, \tilde{X}_N = x_N\right). \tag{3.17}$$

We have the convergence

$$-\frac{1}{N} \log p(x_1, x_2, \ldots, x_N) \to H\left(\tilde{X}\right) \tag{3.18}$$

in probability.

Proof Since the random variables $\tilde{X}_1$, $\tilde{X}_2, \ldots, \tilde{X}_N$ are independent of each other, then we have

$$p(x_1, x_2, \ldots, x_N) = p(x_1)p(x_2)\cdots p(x_N). \tag{3.19}$$

When (3.19) is substituted into

$$-\frac{1}{N}\log p(x_1, x_2, \ldots, x_N)$$

we obtain

$$-\frac{1}{N}\log p(x_1)p(x_2)\cdots p(x_N)$$

which can be written as

$$-\frac{1}{N}\sum_i \log p(x_i). \tag{3.20}$$

According to the weak of large numbers, we have

$$-\frac{1}{N}\sum_i \log p(x_i) = -E\big(\log p\big(\tilde{X}\big)\big)$$

which is nothing but $H(\tilde{X})$, i.e.,

$$-\frac{1}{N}\sum_i \log p(x_i) = H\big(\tilde{X}\big).$$

3.3.1 *Typical Sequences and Typical Set*

We will denote the real number sequence $x_1, x_2, \ldots, x_N$ by x_1^N, i.e.

$$x_1^N = [x_1, x_2, \ldots, x_N]. \tag{3.21}$$

Similarly, the sequence of random variables $\tilde{X}_1$, $\tilde{X}_2, \ldots, \tilde{X}_N$ will be denoted by $\tilde{X}_1^N$. i.e.,

$$\tilde{X}_1^N = \big[\tilde{X}_1, \tilde{X}_2, \ldots, \tilde{X}_N\big]. \tag{3.22}$$

Typical Sequences:

Assume that we have a sequence of real numbers x_1^N produced by sequence of identically distributed random variables $\tilde{X}_1^N$. The sequence x_1^N is accepted as a typical sequence if it satisfies

$$2^{-N\left(H(\tilde{X})+\epsilon\right)} \leq p(x_1, x_2, \ldots, x_N) \leq 2^{-N\left(H(\tilde{X})-\epsilon\right)}. \tag{3.23}$$

Typical Set:

The typical set T_ϵ^N is the set consisting of the typical sequences x_1^N produced by the sequence of identically distributed random variables $\tilde{X}_1^N$.

Example 3.10 What does x_1^4 mean?

Solution 3.10 x_1^4 denoted the real number vector $[x_1\, x_2\, x_3\, x_4]$, i.e.,

$$x_1^4 = [x_1\, x_2\, x_3\, x_4].$$

Example 3.11 What does $Prob\left(\tilde{X}_1^5 = x_1^5\right)$ mean?

Solution 3.11 $Prob\left(\tilde{X}_1^5 = x_1^5\right)$ means

$$Prob\left(\tilde{X}_1 = x_1,\, \tilde{X}_2 = x_2,\, \tilde{X}_3 = x_3,\, \tilde{X}_4 = x_4,\, \tilde{X}_5 = x_5\right)$$

which is nothing but the probability mass function

$$p(x_1,\, x_2,\, x_3,\, x_4,\, x_5).$$

Example 3.12 What does $x_1^N \in T_\epsilon^N$ mean?

Solution 3.12 x_1^N is a real number sequence, i.e., it is a real number vector, T_ϵ^N is the set consisting of real number vectors satisfying typical sequence property. $x_1^N \in T_\epsilon^N$ means that x_1^N is a typical sequence belonging to the typical set T_ϵ^N.

Example 3.13 We have IID random variable sequence

$$\tilde{X}_1,\, \tilde{X}_2, \ldots, \tilde{X}_N \sim p(x)$$

and range set $R_{\tilde{X}} = \{a,\, b,\, c\}$. The probability mass function $p(x)$ is defined on the range set as

$$p(x=a)=\frac{1}{4} \quad p(x=b)=\frac{2}{4} \quad p(x=c)=\frac{1}{4}$$

and the generic random variable having the same probability mass function is indicated as

$$\tilde{X} \sim p(x).$$

(a) For $N = 2$, find the typical sequences, and then find the typical set. How many typical sequences are available in the typical set? Find also the number of non-typical sequences. Calculate the probability of the typical set. Take $\epsilon = 0.1$.

(b) For $N = 3$, find the typical sequences, and then find the typical set. How many typical sequences are available in the typical set? Find also the number of non-typical sequences. Calculate the probability of the typical set. Take $\epsilon = 0.2$.

Solution 3.13

(a) Let's first calculate the entropy of the generic random variable $\tilde{X}$ as in

$$H(\tilde{X}) = -\sum_x p(x)\log p(x) \to H(\tilde{X}) = -\left(\tfrac{1}{4}\log\tfrac{1}{4} + \tfrac{2}{4}\log\tfrac{2}{4} + \tfrac{1}{4}\log\tfrac{1}{4}\right) \to$$
$$\to H(\tilde{X}) = 1.5\,\text{bits/sample.} \tag{3.24}$$

The criteria for the sequence x_1^N to be a typical sequence is

$$2^{-N\left(H(\tilde{X})+\epsilon\right)} \le p(x_1, x_2, \ldots, x_N) \le 2^{-N\left(H(\tilde{X})-\epsilon\right)}$$

in which substituting $N = 2,\ \epsilon = 0.1$ and $H(\tilde{X}) = 1.5$ we get

$$2^{-2(1.5+0.1)} \le p(x_1, x_2) \le 2^{-2(1.5-0.1)}$$

which is simplified as

$$0.1088 \le p(x_1, x_2) \le 0.1436. \tag{3.25}$$

Since the random variables are IID, we can use $p(x_1, x_2) = p(x_1)p(x_2)$ in (3.25), then we get

$$0.1088 \le p(x_1)p(x_2) \le 0.1436. \tag{3.26}$$

Considering (3.26), we can determine the typical sequences as in Table 3.1. Collecting the typical sequences given in Table 3.1, we form the typical set $T_{\epsilon=0.1}^{N=2}$ as in

$$T_{0.1}^2 = \{ab,\ ba,\ bc,\ cb\}$$

where it is seen that we have 4 typical sequences. The number of sequences for $N = 2$ equals to $3^2 = 9$, i.e., $aa,\ bb,\ cc,\ ab,\ ba,\ ac,\ ca,\ bc,\ cb$. The number of non-typical sequences in this case equals to $9 - 4 = 5$. We see that the number of typical sequences is less than the number of non-typical sequences.
The probability of the typical set can be calculated as

$$Prob\left(T_{0.1}^2\right) = Prob(ab,\ ba,\ bc,\ cb) = Prob(ab) + Prob(ba) + Prob(bc) + Prob(cb) \rightarrow$$
$$Prob\left(T_{0.1}^2\right) = 0.5.$$

We see that although typical set includes less number of sequences, the probability of the set it high, i.e., it gets half of the all probabilities.

(b) Substituting $N = 3$, $\epsilon = 0.2$ and $H\left(\tilde{X}\right) = 1.5$ into

$$2^{-N\left(H\left(\tilde{X}\right)+\epsilon\right)} \le p(x_1, x_2, \ldots, x_N) \le 2^{-N\left(H\left(\tilde{X}\right)-\epsilon\right)}$$

we get

$$2^{-3(1.5+0.2)} \le p(x_1, x_2, x_3) \le 2^{-3(1.5-0.2)}$$

which is simplified as

$$0.0292 \le p(x_1, x_2, x_3) \le 0.0670$$

in which using $p(x_1, x_2, x_3) = p(x_1)p(x_2)p(x_3)$ we get

$$0.0292 \le p(x_1)p(x_2)p(x_3) \le 0.0670. \tag{3.27}$$

Considering (3.27), we can determine the typical sequences as in Table 3.2.

Table 3.1 Typical sequences for $N = 2$

x_1	x_2	$p(x_1)p(x_2)$
a	b	$p(a)p(b) = 0.1250$
b	a	$p(b)p(a) = 0.1250$
b	c	$p(b)p(c) = 0.1250$
c	b	$p(c)p(b) = 0.1250$

Table 3.2 Typical sequences for $N = 3$

x_1	x_2	x_3	$p(x_1)p(x_2)p(x_3)$
a	b	c	$p(a)p(b)p(c) = 0.0313$
a	c	b	$p(a)p(c)p(b) = 0.0313$
b	a	c	$p(b)p(a)p(c) = 0.0313$
b	c	a	$p(b)p(c)p(a) = 0.0313$
c	a	b	$p(c)p(a)p(b) = 0.0313$
c	b	a	$p(c)p(b)p(a) = 0.0313$
a	a	b	$p(a)p(a)p(b) = 0.0313$
a	b	a	$p(a)p(b)p(a) = 0.0313$
b	a	a	$p(b)p(a)p(a) = 0.0313$
c	c	b	$p(c)p(c)p(b) = 0.0313$
c	b	c	$p(c)p(b)p(c) = 0.0313$
b	c	c	$p(b)p(c)p(c) = 0.0313$

When the Table 3.2 is inspected we see that there are 12 typical sequences, and the typical set for $N = 3$, and $\epsilon = 0.2$ can be formed as

$$T_{0.2}^3 = \{abc,\ acb,\ bac,\ bca,\ cab,\ cba,\ aab,\ aba,\ baa,\ ccb,\ cbc,\ bcc\}.$$

The probability of the typical set equals to

$$Prob\left(T_{0.2}^3\right) = 12 \times \frac{1}{16} \rightarrow Prob\left(T_{0.2}^3\right) = 0.75.$$

The total number of sequences is $3^3 = 27$. The number of non-typical sequences is $27 - 12 = 15$. We see that we have less number of typical sequences than the number of non-typical sequences, but the probability of the typical set, 0.75 in this case, is much higher than the probability of non-typical set which can be calculated as $1 - 0.75 = 0.25$.

Note: In this example, the random variables are IID. For this reason, we wrote in part-a $p(x_1, x_2) = p(x_1)p(x_2)$. However, please keep in your mind that this may not be like this in every case. If the random variables are not IID, then we need to use joint probability mass function directly in

$$2^{-N\left(H(\tilde{X})+\epsilon\right)} \leq p(x_1, x_2, \ldots, x_N) \leq 2^{-N\left(H(\tilde{X})-\epsilon\right)}$$

to determine the typical sequences.

These examples are given for illustrative purposes only. In fact, in reality we consider sequences having very long lengths.

3.3.2 Strongly and Weakly Typical Sequences

The typical sequences can be divided into two classes. These classes are strongly typical sequences and weakly typical sequences. In both classes, typical sequences satisfy the criteria

$$2^{-N\left(H(\tilde{X})+\epsilon\right)} \leq p(x_1, x_2, \ldots, x_N) \leq 2^{-N\left(H(\tilde{X})-\epsilon\right)}. \tag{3.28}$$

Let the generic random variable $\tilde{X}$ has the range set $R_{\tilde{X}} = \{x_1, x_2, \ldots, x_r\}$, and probability mass function $p(x)$ be given as

$$p(x_1) = p_1 \quad p(x_2) = p_2 \quad p(x_3) = p_3 \cdots p(x_r) = p_r$$

In a strongly typical sequence x_1^N, the symbol x_1 appears in x_1^N sequence $N \times p_1$ times, the symbol x_2 appears in x_1^N sequence $N \times p_2$ times, and son on, till the last symbol x_r which appears in x_1^N sequence $N \times p_r$ times.
In a weakly typical sequence, we do not have a constraint like the one for strongly typical sequences. The only constraint is the (3.28) for a weakly typical sequence. When we say that a sequence is typical, we do not classify it as strongly or weakly, it may be a strongly or weakly typical sequence.

Properties of Typical Sequences

Assume that x_1^N is a typical sequence belonging to the typical set T_ϵ^N, i.e., $x_1^N \in T_\epsilon^N$. Then, we have the following properties about the given typical sequence and typical set.

Theorem 3.1 *If* $x_1^N \in T_\epsilon^N$, *then we have*

$$H(\tilde{X}) - \epsilon \leq -\frac{1}{N} p(x_1, x_2, \ldots, x_N) \leq H(\tilde{X}) + \epsilon. \tag{3.29}$$

Proof 3.1 $x_1^N \in T_\epsilon^N$ implies that

$$2^{-N\left(H(\tilde{X})+\epsilon\right)} \leq p(x_1, x_2, \ldots, x_N) \leq 2^{-N\left(H(\tilde{X})-\epsilon\right)}$$

in which taking the $\log(\cdot)$ of the three parts and rearranging, we get

$$H(\tilde{X}) - \epsilon \leq -\frac{1}{N} p(x_1, x_2, \ldots, x_N) \leq H(\tilde{X}) + \epsilon.$$

which can be written in a more compact manner as

$$\left| -\frac{1}{N}p(x_1, x_2, \ldots, x_N) - H(\tilde{X}) \right| \leq \epsilon.$$

In our Example 3.13, we have seen that as the length of the sequence increases, the probability of the typical set increases as well.

Theorem 3.2 *If T_ϵ^N is a typical set, then we have*

$$Prob\left(T_\epsilon^N\right) > 1 - \epsilon \tag{3.30}$$

where N is a sufficiently large positive integer.

Proof 3.2 From asymptotic equipartition theorem, we have

$$-\frac{1}{N}\log p(x_1, x_2, \ldots, x_N) \xrightarrow{p} H(\tilde{X}). \tag{3.31}$$

Considering the definition of convergence in probability, (3.31) implies that

$$\lim_{N\to\infty} Prob\left(\left| -\frac{1}{N}\log p(x_1, x_2, \ldots, x_N) - H(\tilde{X}) \right| > \epsilon\right) = 0. \tag{3.32}$$

There exists and N_0 such that $N \geq N_0$ for which (3.32) can be written as

$$Prob\left(\underbrace{x_1^N;\ \left| -\frac{1}{N}\log p(x_1, x_2, \ldots, x_N) - H(\tilde{X}) \right| < \epsilon_1}_{\substack{\text{means the set of } x_1^N \text{ satisfying} \\ \left| -\frac{1}{N}\log p(x_1, x_2, \ldots, x_N) - H(\tilde{X}) \right| < \epsilon_1}}\right) > 1 - \epsilon_2 \tag{3.33}$$

which has the same meaning as

$$Prob\left(T_\epsilon^N\right) > 1 - \epsilon. \tag{3.34}$$

Theorem 3.3 $\left|T_\epsilon^N\right|$ *which indicates the number of elements in the typical set T_ϵ^N satisfy*

$$\left|T_\epsilon^N\right| \leq 2^{n\left(H(\tilde{X}) + \epsilon\right)}. \tag{3.35}$$

Proof 3.3 Let T_x^N be the set of all sequences of length N, i.e., the set of typical and non-typical sequences together. Then, it is obvious that

$$\sum_{x_1^N \in T_x^N} p(x_1^N) = 1 \tag{3.36}$$

which means that sum of the probabilities of the all sequences equals to 1. If we consider only the typical sequences rather than all the sequences, then we can write,

$$\sum_{x_1^N \in T_\epsilon^N} p(x_1^N) \leq 1 \tag{3.37}$$

in which using the property

$$2^{-N\left(H(\tilde{X})+\epsilon\right)} \leq p\left(x_1^N\right)$$

for the typical sequence x_1^N, we obtain

$$\sum_{x_1^N \in T_\epsilon^N} 2^{-N\left(H(\tilde{X})+\epsilon\right)} \leq 1$$

which can be written as

$$2^{-N\left(H(\tilde{X})+\epsilon\right)} \left(\sum_{x_1^N \in T_\epsilon^N} 1 \right) \leq 1 \tag{3.38}$$

where

$$\sum_{x_1^N \in T_\epsilon^N} 1 = \left|T_\epsilon^N\right|$$

i.e., the number of elements in the typical set T_ϵ^N. Then, (3.38) is written as

$$2^{-N\left(H(\tilde{X})+\epsilon\right)} \left|T_\epsilon^N\right| \leq 1$$

which leads to the inequality

$$\left|T_\epsilon^N\right| \leq 2^{N\left(H(\tilde{X})+\epsilon\right)}.$$

Theorem 3.4 *The number of elements in the typical set satisfies*

$$\left|T_\epsilon^N\right| \geq (1-\epsilon)2^{n\left(H\left(\tilde{X}\right)-\epsilon\right)}. \tag{3.39}$$

Proof 3.4 For a sufficiently large N value, the probability of the typical set satisfy

$$Prob\left(T_\epsilon^N\right) \geq 1-\epsilon$$

which can be written as

$$\sum_{x_1^N \in T_\epsilon^N} p\left(x_1^N\right) \geq 1-\epsilon$$

where employing

$$2^{-N\left(H\left(\tilde{X}\right)-\epsilon\right)} \geq p\left(x_1^N\right)$$

we get

$$\sum_{x_1^N \in T_\epsilon^N} 2^{-N\left(H\left(\tilde{X}\right)-\epsilon\right)} \geq 1-\epsilon$$

which can be written as

$$2^{-N\left(H\left(\tilde{X}\right)-\epsilon\right)} \underbrace{\sum_{x_1^N \in T_\epsilon^N} 1}_{T_\epsilon^N} \geq 1-\epsilon$$

which is simplified as

$$\left|T_\epsilon^N\right| \geq (1-\epsilon)2^{N\left(H\left(\tilde{X}\right)-\epsilon\right)}.$$

Example 3.14 What is the difference between T_ϵ^N and T_x^N?

Solution 3.14 T_ϵ^N denotes the typical set consisting of typical sequence which consists of N symbols, on the other hand, T_x^N denotes the set consisting of sequences consisting of N symbols, these sequences can both be typical and non-typical, i.e., all types of sequences.

Example 3.15 What is the difference between T_ϵ^3 and T_ϵ^5?

Solution 3.15 T_ϵ^3 is the typical set consisting of typical sequences of length 3. For example,

$$T_\epsilon^3 = \{aba,\ bcb,\ cac,\ dad, \ldots\}.$$

T_ϵ^5 is the typical set consisting of typical sequences of length 5. For example,

$$T_\epsilon^5 = \{abaca,\ bcbca,\ cacdb,\ ddadb, \ldots\}.$$

Example 3.16 T_ϵ^N is a typical set and N is a large number. Comment on the probability of the typical set, i.e., what can be $Prob\left(T_\epsilon^N\right) = ?$

Solution 3.16 According to Theorem 3.2 we have,

$$Prob\left(T_\epsilon^N\right) > 1 - \epsilon$$

which means that $Prob\left(T_\epsilon^N\right)$ is a number close to 1. For instance, $Prob\left(T_\epsilon^N\right)$ can be equal to 0.99, i.e., $Prob\left(T_\epsilon^N\right) = 0.99$.

Example 3.17 Is there a way to find the number of typical sequences in a typical set?

Solution 3.17 According to Theorems 3.3 and 3.4, the number of elements in a typical set satisfies

$$\left|T_\epsilon^N\right| \leq 2^{N\left(H\left(\tilde{X}\right)+\epsilon\right)}$$

and

$$\left|T_\epsilon^N\right| \geq (1-\epsilon)2^{N\left(H\left(\tilde{X}\right)-\epsilon\right)}.$$

Combining both inequalities, we obtain

$$(1-\epsilon)2^{N(H(\epsilon)-\epsilon)} \leq \left|T_\epsilon^N\right| \leq 2^{N\left(H\left(\tilde{X}\right)+\epsilon\right)}. \tag{3.40}$$

As $\epsilon \rightarrow 0$, (3.40) reduces to

$$2^{NH\left(\tilde{X}\right)} \leq \left|T_\epsilon^N\right| \leq 2^{NH\left(\tilde{X}\right)}$$

which means that for very large N values, the number of elements in a typical set equals to $2^{NH(\tilde{X})}$, i.e.,

$$\left|T_{\epsilon}^{N}\right| \approx 2^{NH(\tilde{X})}. \tag{3.41}$$

If the number of elements in a typical set equals to $\left|T_{\epsilon}^{N}\right|$, then the probability of a typical sequence equals to

$$p\left(x_1^N\right) \approx \frac{1}{\left|T_{\epsilon}^{N}\right|}$$

where employing (3.41), we get

$$p\left(x_1^N\right) \approx 2^{-NH(\tilde{X})}. \tag{3.42}$$

Example 3.18 Let $\tilde{X}_1, \tilde{X}_2, \ldots, \tilde{X}_N \sim p(x)$, and $\tilde{X} \sim p(x)$. The range set for $\tilde{X}$ is defined as

$$R_{\tilde{X}} = \{d, e, f\}.$$

Write some sample sequences for x_1^N where $N = 6$ and $N = 13$

Solution 3.18 x_1^6 is a vector consisting of 6 symbols which are chosen from $R_{\tilde{X}}$. We can write the following sample sequences for x_1^6

$$\begin{gathered} dedfdf \\ eedefd \\ ffedef \\ \vdots \end{gathered}$$

Similarly, x_1^{13} is a vector consisting of 13 symbols which are chosen from $R_{\tilde{X}}$. We can write the following sample sequences for x_1^{13}

$$\begin{gathered} ddefdfefdfedf \\ ffedfdedffedf \\ eefdfedfdefde \\ \vdots \end{gathered}$$

Example 3.19 Let $\tilde{X}_1, \tilde{X}_2, \ldots, \tilde{X}_N \sim p(x)$, and $\tilde{X} \sim p(x)$. The range set for $\tilde{X}$ and probability mass function $p(x)$ is defined as

$$R_{\tilde{X}} = \{a, b, c\}$$
$$p(x=a) = \tfrac{1}{4}, \quad p(x=b) = \tfrac{2}{4}, \quad p(x=c) = \tfrac{1}{4}.$$

Find the number of typical sequences for $N = 20$, and $N = 50$.

Solution 3.19 The entropy of the random variable $\tilde{X}$ can be calculated using

$$H(\tilde{X}) = -\sum_x p(x)\log p(x) \rightarrow H(\tilde{X}) = 1.5\,\text{bits/symbol}.$$

Using $|T_\epsilon^N| = 2^{NH(\tilde{X})}$, we find the number of typical sequences for $N = 20$, and $N = 50$ as

$$|T_\epsilon^{20}| = 2^{30}, \qquad |T_\epsilon^{50}| = 2^{75}.$$

An Alternative Approach to Find the Number of Typical Sequences in a Typical Set

Let $\tilde{X}_1, \tilde{X}_2, \ldots, \tilde{X}_N \sim p(x)$, and $\tilde{X} \sim p(x)$. The range set for $\tilde{X}$ and probability mass function $p(x)$ is defined as

$$R_{\tilde{X}} = \{s_1, s_2, \ldots, s_r\}$$
$$p(x=s_1) = p_1 \quad p(x=s_2) = p_2 \cdots p(x=s_r) = p_r.$$

As $N \rightarrow \infty$, in the symbol vector

$$x_1^N = [x_1\, x_2 \cdots x_N]$$

the symbol s_1 is repeated Np_1 times, the symbol s_2 is repeated Np_2 times, and so on, till the symbol s_r which is repeated Np_r times.
Then, the probability of the sequence x_1^N can be calculated as

$$\begin{aligned} p(x_1, x_2, \ldots, x_N) &= p_1^{Np_1} \times p_2^{Np_2} \times \cdots \times p_r^{Np_r} \\ &= 2^{\log p_1^{Np_1}} \times 2^{\log p_2^{Np_2}} \times \cdots \times 2^{\log p_r^{Np_r}} \\ &= 2^{Np_1 \log p_1} \times 2^{Np_2 \log p_2} \times \cdots \times 2^{Np_r \log p_r} \\ &= 2^{N\sum_i p_i \log p_i} \\ &= 2^{-NH(\tilde{X})}. \end{aligned} \tag{3.43}$$

which is the same result as (3.42). And the number of typical sequences can be obtained from (3.43) as

$$\left|T_{\epsilon}^{N}\right| = 2^{NH(\tilde{X})}.$$

Note: $p(x_1, x_2, \ldots, x_N) = p(x_1^N) = Prob(\tilde{X}_1^N = x_1^N)$. While deriving (3.43), we started with the definition of strongly typical sequences. However, the obtained result includes the number of all types of typical sequences. This is due to the combination of same probabilities representing different symbols.

Example 3.20 For the random variable $\tilde{X}$, the range set and probability mass function are given as

$$R_{\tilde{X}} = \{a, b, c\}$$
$$p(x = a) = \tfrac{1}{4} \quad p(x = b) = \tfrac{2}{4} \quad p(x = c) = \tfrac{1}{4}.$$

(a) Let $\tilde{X}_1^N \sim p(x)$. Find a strongly typical sequence x_1^N for $N = 20$, and verify the inequality

$$2^{-N(H(\tilde{X})+\epsilon)} \leq p(x_1, x_2, \ldots, x_N) \leq 2^{-N(H(\tilde{X})-\epsilon)} \tag{3.44}$$

for your sequence.

(b) Find the total number of strongly typical sequences, and weakly typical sequences.

Solution 3.20

(a) In a strongly typical sequence x_1^{20}, the symbol 'a' appears $20 \times \frac{1}{4} = 5$, the symbol '$b$' appears $20 \times \frac{2}{4} = 10$ times, and the symbol 'c' appears $20 \times \frac{1}{4} = 5$ times. Then, we can form a strongly typical sequence as

$$x_1^{20} = [ababababab bcbcbcbcbc]. \tag{3.45}$$

The entropy of the generic random variable $\tilde{X} \sim p(x)$ can be calculated as

$$H(\tilde{X}) = 1.5.$$

Let's choose $\epsilon = 0.01$, then the inequality (3.44) for the found sequence in (3.45) happens to be as

$$2^{-20(1.5+0.01)} \leq \underbrace{p(x_1, x_2, \ldots, x_N)}_{\left(\frac{1}{4}\right)^5 \left(\frac{2}{4}\right)^{10} \left(\frac{1}{4}\right)^5} \leq 2^{-20(1.5-0.01)}$$

leading to

$$8.108 \times 10^{-10} \leq 9.32 \times 10^{-10} \leq 1.07 \times 10^{-9} \; \surd$$

which is a correct inequality.
(b) The total number of strongly typical sequences can be found using

$$\binom{20}{5}\binom{15}{10}\binom{5}{5} \rightarrow 46558512.$$

The total number of typical sequences can be calculated as

$$2^{NH(\tilde{X})} \rightarrow 2^{30} \rightarrow 1.0737 \times 10^{9}.$$

Then, the total number of weakly typical sequences is

$$1.0737 \times 10^{9} - 46558512 = 1.0272 \times 10^{9}$$

from which we see that, most of the typical sequences fall into the category of weakly typical sequences.

Example 3.21 Let $\tilde{X}_1, \tilde{X}_2, \ldots, \tilde{X}_N \sim p(x)$, and $\tilde{X} \sim p(x)$. The range set for $\tilde{X}$ is defined as

$$R_{\tilde{X}} = \{a, b, c\}.$$

Find the total number of sequences x_1^N for $N = 10$.

Solution 3.21 The total number of sequences x_1^N equals to $\left|R_{\tilde{X}}\right|^N$. For our example $\left|R_{\tilde{X}}\right| = 3$, and the total number of sequences x_1^{10} equals to 3^{10}.

Example 3.22 Let $\tilde{X}_1, \tilde{X}_2, \ldots, \tilde{X}_N \sim p(x)$, and $\tilde{X} \sim p(x)$. The range set for $\tilde{X}$ and probability mass function $p(x)$ is defined as

$$R_{\tilde{X}} = \{a, b, c\}$$
$$p(x = a) = \tfrac{1}{4} \quad p(x = b) = \tfrac{2}{4} \quad p(x = c) = \tfrac{1}{4}.$$

Find the number of typical and non-typical sequences for $N = 10$.

Solution 3.22 The entropy of the random variable $\tilde{X}$ can be calculated as $H(\tilde{X}) = 1.5$. The number of all sequences x_1^{10} equals to $\left|R_{\tilde{X}}\right|^{10} = 3^{10}$. The number of typical sequences equals to $2^{NH(\tilde{X})} = 2^{15}$. The number of non-typical sequences equals to $3^{10} - 2^{15}$.

Representation of Symbol Sequences by Bit Vectors:

In communication engineering, information is represented by real number sequences. But for the transmission of real number sequences, we need to represent these sequences by bit vectors. A total number of K different real number sequences can be represented by bit vectors each having $\log K$ bits, where $\lceil \cdot \rceil$ is ceil operator.

The total number of typical sequences equals to $2^{NH(\tilde{X})}$, and each typical sequence can be represented by a bit vector having

$$\left\lceil \log 2^{NH(\tilde{X})} \right\rceil = \lceil NH(\tilde{X}) \rceil \tag{3.46}$$

number of bits.

The total number of non-typical sequences equals to $|R_{\tilde{X}}|^N - 2^{NH(\tilde{X})}$, and each non-typical sequence can be represented by a bit vector having

$$\left\lceil \log\left(|R_{\tilde{X}}|^N - 2^{NH(\tilde{X})} \right) \right\rceil$$

number of bits.

On the other hand, without classifying as typical or non-typical any sequence x_1^N can be represented by

$$\left\lceil \log |R_{\tilde{X}}|^N \right\rceil = \lceil N \log |R_{\tilde{X}}| \rceil \tag{3.47}$$

number of bits.

However, for large even moderate values of N, the number of typical sequences becomes very small compared to the number of non-typical sequences, and the non-typical sequences can also be represented by

$$\left\lceil \log |R_{\tilde{X}}|^N \right\rceil = \lceil N \log |R_{\tilde{X}}| \rceil \tag{3.48}$$

number of bits.

If we want to distinguish typical sequences from non-typical sequences, we can prefix the binary representation of typical sequences by bit ‘0’, and prefix the binary representation of non-typical sequences by ‘1’. In this case, we need

$$\left\lceil \log 2^{NH(\tilde{X})} \right\rceil + 1 = \lceil NH(\tilde{X}) \rceil + 1 \tag{3.49}$$

bits to represent a typical sequence, on the other hand, we need

$$\left\lceil \log \left| R_{\tilde{X}} \right|^N \right\rceil + 1 = \left\lceil N \log \left| R_{\tilde{X}} \right| \right\rceil + 1 \tag{3.50}$$

bits to represent a non-typical sequence.

Example 3.23 Let $\tilde{X}_1, \tilde{X}_2, \ldots, \tilde{X}_N \sim p(x)$, and $\tilde{X} \sim p(x)$. The range set for $\tilde{X}$ is defined as

$$R_{\tilde{X}} = \{a, b, c\}.$$

Find the total number of sequences x_1^N for $N = 5$. We want to represent each sequence by a bit vector. How many bits do we need to use for each vector for this purpose?

Solution 3.23 The total number of sequences is $\left| R_{\tilde{X}} \right|^N = 3^5 = 243$. To represent each sequence by a bit vector we need $\lceil \log 243 = 7.92 \rightarrow 8\,\text{bits} \rceil$ in each vector. An example of this representation is shown in Table 3.3.

Example 3.24 Let $\tilde{X}_1, \tilde{X}_2, \ldots, \tilde{X}_N \sim p(x)$, and $\tilde{X} \sim p(x)$. The range set for $\tilde{X}$ and probability mass function $p(x)$ is defined as

$$R_{\tilde{X}} = \{a, b, c\}$$
$$p(x = a) = \tfrac{1}{16} \quad p(x = b) = \tfrac{7}{8} \quad p(x = c) = \tfrac{1}{16}.$$

Find the number of typical and non-typical sequences for $N = 20$. How many bits are required to represent a typical sequence?

Solution 3.24 The entropy of the random variable $\tilde{X}$ can be calculated as

$$H(\tilde{X}) = -\sum_x p(x) \log p(x) \rightarrow H(\tilde{X}) = -\left(\frac{1}{16} \log \frac{1}{16} + \frac{7}{8} \log \frac{7}{8} + \frac{1}{16} \log \frac{1}{16} \right) \rightarrow$$

$$H(\tilde{X}) = 0.6686\,\text{bits/symbol}.$$

Table 3.3 Bit vector representation of sequences

Sequence x_1^5	Bit vector representation
aaaaa	00000000
aaaab	00000001
aaaba	00000010
…	…
ccccc	11110011

The number of typical sequences is $2^{NH(\tilde{X})} = 2^{20\times 0.6686} \to 10543$. The number of all sequences x_1^{20} equals to $3^{20} = 3486800000$. The number of non-typical sequences is $3^{20} - 2^{20\times 0.6686} \approx 3^{20}$. As we see, the number of typical sequences is very small when it is compared to the number of non-typical sequences.

The probability of a typical sequence equals to $1/10543$ and the probability of a non-typical sequence equals to $1/3486800000$. It is obvious that the probability of a typical sequence is much more than the probability of a non-typical sequence even for $N = 20$.

In fact, for large N values non-typical sequences never appear, since their occurrence probability becomes almost equal to zero. Typical sequences become small in number but they get all the probability.

A typical sequence can be represented by $\lceil NH(\tilde{X}) \rceil = \lceil 20 \times 0.6686 \rceil = \lceil 13.37 \rceil \to 14$ bits.

A non-typical sequence can be represented by $\lceil N \log |R_{\tilde{X}}| \rceil = \lceil 20 \log 3 \rceil = \lceil 31.69 \rceil \to 32$ bits.

3.4 Data Compression or Source Coding

The binary vectors representing the real number sequences are considered as code-words. And the mapping process between real number sequences and binary vectors is called source encoding. Assume that we made mapping for all the real number sequences, then we wonder how many bits in average we used for the binary vectors. As this average number gets smaller we better utilize the resources we have. Now, let's consider the average bit-vector length used for source encoding operation.

Average Bit-Vector Length (Average Code-Word Length):

Assume that we transmit a number of bit-vectors, i.e., code-words, representing some real number sequences. In this case, we should consider the probabilistic average. Since, it is obvious that we don't need to transmit all the available sequences. For the real number sequence x_1^N, let c_x be the code-word used, and let $L(c_x)$ be the length of the code-word, i.e., the number of bits available in the code-word.

The expected length of the code-words can be calculated using

$$E(L(\tilde{C}_x)) = \sum_{x_1^N} p(x_1^N) L(c_x) \tag{3.51}$$

which can be written as the sum of two parts as in

$$E\left(L\left(\tilde{C}_x\right)\right)=\underbrace{\sum_{x_\epsilon^N}}_{\substack{Typical\\Sequences}} p\left(x_1^N\right)\underbrace{L(c_x)}_{NH}+\underbrace{\sum_{{}^c x_\epsilon^N}}_{\substack{NonTypical\\Sequences}} p\underbrace{\left(x_1^N\right)}_{\approx 0}L(c_x)$$

which is simplified as

$$E\left(L\left(\tilde{C}_x\right)\right)\approx NH\left(\tilde{X}\right)\underbrace{\underbrace{\sum_{x_\epsilon^N}}_{\substack{Typical\\Sequences}} p\left(x_1^N\right)}_{=Prob\left(T_\epsilon^N\right)\to 1}\to E\left(L\left(\tilde{C}_x\right)\right)\approx NH\left(\tilde{X}\right)$$

Thus,

$$E\left(L\left(\tilde{C}_x\right)\right)\approx NH\left(\tilde{X}\right) \tag{3.52}$$

which means that we can represent sequences x_1^N using $NH\left(\tilde{X}\right)$ bits on average. In fact, this is the main idea of the data compression topic. The formula in (3.52) can also be written as

$$E\left(L\left(\frac{\tilde{C}_x}{N}\right)\right)\approx H\left(\tilde{X}\right) \tag{3.53}$$

which shows the number of bits used on average to represent a single symbol appearing in the sequence x_1^N.

We indicated in (3.53) that it is possible to represent a single symbol appearing in real number sequences by $H\left(\tilde{X}\right)$ number of bits. This is the theoretical result we found. Next, the engineering part comes. How should we do the mapping between real number sequences and bit-vectors? In other words, how should we find source code-words such that the number of bits used per symbol approaches $H\left(\tilde{X}\right)$. This problem leads to a new engineering field called source code design. For years, engineers are working on this problem, and they are trying to find better source encoding methods compared to the methods found in past.

Data Compression or Source Encoding

A source code C is nothing but a set, whose elements are symbol vectors which can be of different lengths.

Since in communication engineering bits are employed to convey information, we will focus on sources involving bit-vectors only.

A binary source code C is nothing but a set whose elements are bit vectors, which can be of different lengths. Since we will only use binary sources in this book, we will use the term "source code" for "binary source code".

Example 3.25 A source code C is defined as

$$C = \{0000,\ 001,\ 01, 1\}.$$

Source encoding, or data compression, is nothing but a mapping between either symbols of range set of a generic random variable and bit vectors, or it is a mapping between real number sequences and bit vectors.
We will use $c(x)$ to denote the binary vector, i.e., code-word, mapped to symbol x appearing in the range set $R_{\tilde{X}}$. The length of $c(x)$, i.e., number of bits in $c(x)$, will be indicated as $l(c(x))$.
In a similar manner, we will use $c\left(x_1^N\right)$ to denote the binary vector, i.e., code-word, mapped to sequence x_1^N. The length of $c(x_1^N)$, i.e., number of bits in $c(x_1^N)$, will be indicated as $l(c\left(x_1^N\right))$ also.

Example 3.26 Let $\tilde{X}$ be a discrete random variable whose range set is $R_{\tilde{X}} = \{a,\ b,\ c,\ d\}$. The code-words $c(x)$ for the symbols in the range set is defined as

$$c(a) = 00, \quad c(b) = 010, \quad c(c) = 11, \quad c(d) = 110.$$

And the source code C can be indicated as

$$C = \{00,\ 010,\ 11,\ 110\}.$$

The lengths of the code-words are

$$l(c(a)) = 2, \quad l(c(b)) = 3, \quad l(c(c)) = 2, \quad l(c(d)) = 3.$$

We can use $l(c_x)$ instead of $l(c(x))$ for the simplicity of the notation in our formulas.

Definition The probabilistic average length $L(C)$ of a source code C for the symbols of the range set of a random variable $\tilde{X}$ with probability mass function $p(x)$ is calculated by

$$L(C) = \sum_{x \in R_{\tilde{X}}} p(x) l(c_x) \tag{3.54}$$

where $l(c_x)$ is the length of the code-word associated with the symbol x.

Example 3.27 For the discrete random variable $\tilde{X}$, the range set $R_{\tilde{X}}$, and the probability mass function $p(x)$ are given as

$$R_{\tilde{X}} = \{a, b, c, d\},$$
$$p(x = a) = \tfrac{1}{2}, \quad p(x = b) = \tfrac{1}{4}, \quad p(x = c) = \tfrac{1}{8}, \quad p(x = d) = \tfrac{1}{8}.$$

The binary code-words for the symbols in the range set are defined as

$$c(a) = 0, \quad c(b) = 10, \quad c(c) = 110, \quad c(d) = 111.$$

Find the expected value, i.e., probabilistic average, of the code-words.

Solution 3.27 First, let's calculate the lengths of the code-words as follows

$$l(c(a)) = 1, \quad l(c(b)) = 2, \quad l(c(b)) = 3, \quad l(c(d)) = 3.$$

Employing the formula

$$L(C) = \sum_{x \in R_{\tilde{X}}} p(x) l(c(x))$$

for the given random variable and code-words, we get

$$\begin{aligned} L(C) &= p(a)l(c(a)) + p(b)l(c(b)) + p(c)l(c(c)) + p(d)l(c(d)) \\ &= \frac{1}{2} \times 1 + \frac{1}{4} \times 2 + \frac{1}{8} \times 3 + \frac{1}{8} \times 3 \\ &= 1.75 \text{ bits.} \end{aligned}$$

Definition A source code C is said to be non-singular if every symbol in the range set of $\tilde{X}$ is mapped to a different bit-vector (code-word) in C, i.e.,

$$\text{if } x \neq x', \text{ then } c(x) \neq c(x'). \tag{3.55}$$

Example 3.28 Let $R_{\tilde{X}} = \{a, b, c, d\}$, and the code is defined as

$$c(a) = 0, \quad c(b) = 10, \quad c(c) = 110, \quad c(d) = 111.$$

The code C is a non-singular code.

Definition The extension code C^e is used for the sequence of symbols and it is formed as follows

$$c^e\left(x_1^N\right) = c(x_1x_2\cdots x_N) \rightarrow c(x_1x_2\cdots x_N) = c(x_1)c(x_2)\cdots c(x_N) \tag{3.56}$$

where $c(x_1)c(x_2)\cdots c(x_N)$ indicates the concatenation of $c(x_1), c(x_2), \ldots, c(x_N)$, i.e., concatenation of the code-words used for symbols.

Example 3.29 Let $R_{\tilde{X}} = \{a,\ b,\ c,\ d\}$, and the code is defined as

$$c(a) = 0, \quad c(b) = 10, \quad c(c) = 110, \quad c(d) = 111.$$

We can form the following code-words for symbol sequences *aab* and *bcdaab*

$$c^e(aab) = c(a)c(a)c(b) \rightarrow c(aab) = 0010$$
$$c^e(bcdaab) = c(b)c(c)c(d)c(a)c(a)c(b) \rightarrow c^e(abcdaab) = 101101110010.$$

Definition A source code is a uniquely decodable code if its extension code is non-singular.

Example 3.30 Let $R_{\tilde{X}} = \{a,\ b,\ c,\ d\}$, and the code is defined as

$$c(a) = 0, \quad c(b) = 10, \quad c(c) = 110, \quad c(d) = 11.$$

We get the following code-words for the sequences *dab* and *cb*

$$c^e(dab) = 11010 \quad c^e(cb) = 11010.$$

We see that $dab \neq cb$, but $c^e(dab) = c^e(cb)$. For this reason, the code c^e is singular. Hence, the code C is not uniquely decodable.

Definition A prefix code or an instantaneous source code is the one in which no code-word is a prefix of any other code-word.

3.4.1 *Kraft Inequality*

Before studying the Kraft inequality, let's prepare ourselves for the proof of the Kraft inequality. For this purpose, let's introduce the code tree, and study some examples on the code tree.

Tree Branch:

Tree branch is depicted in Fig. 3.1 where each of the left and right edges correspond to a single bit.

Fig. 3.1 Tree branch

Tree:

A tree is formed by concatenating the tree branches from top to bottom manner. A sample tree is depicted in Fig. 3.2.

The connected path from the top node to a lower node refers to a code-word. For instance, formation of a code-word is depicted in Fig. 3.3.

Let l_i denote the index of the levels. For instance, the index of the first level is 0, i.e., $l_0 = 0$. The top most node is called the root node. Any node at level-i, i.e., at level l_i, has $2^{l_j - l_i}$ descendant nodes at level l_j.

Example 3.31 A node at level-1, i.e., at l_1, has $2^{4-1} = 8$ descendant nodes at level-4, i.e., at l_4. This is illustrated in Fig. 3.4.

The prefix codes have the property that no code-word can be the prefix of any other code-word. This condition implies that while forming the code-words using the code tree, if a code-word is formed, its descendant cannot be used to form another code-word.

Fig. 3.2 Tree structure

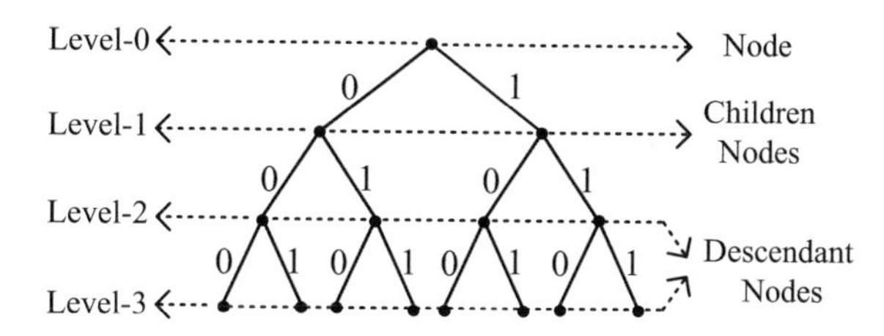

Fig. 3.3 Code-word formation using tree structure

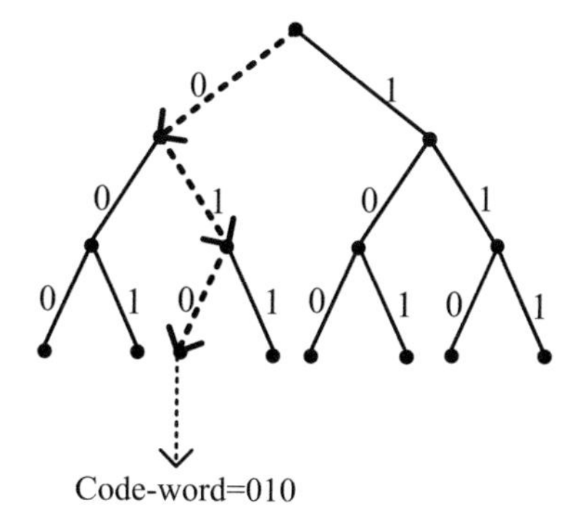

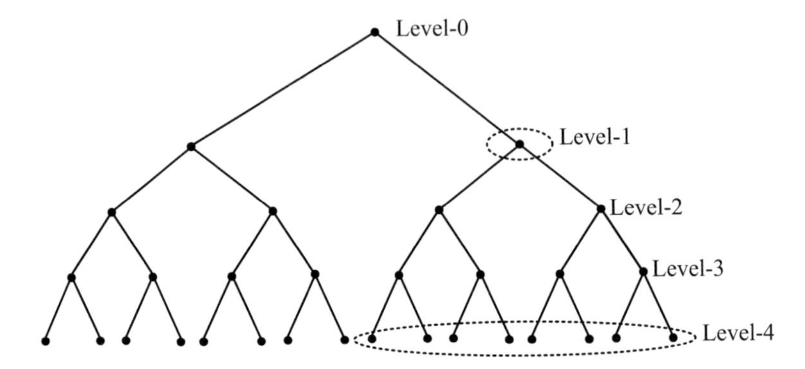

Fig. 3.4 Determination of descendant nodes

Example 3.32 Assume that there are 6 levels in the code tree. How many nodes are available in the bottom most level in this case?

Solution 3.32 If there are 6 levels in the code tree, then the index of the first level is 0 and the index of the last level is 5. In this case, there are

$$2^{5-0} = 32$$

nodes in the last level.

Example 3.33 Let's form a code-word using the dotted path shown in Fig. 3.5. The code-word generated from Fig. 3.5 is '10'. We cannot use the descendant nodes of the destination node of the generated code-word to form any other code-word. This is illustrated in Fig. 3.6.
Descendants of different code-words form disjoint node sets. This issue is illustrated in Fig. 3.7.

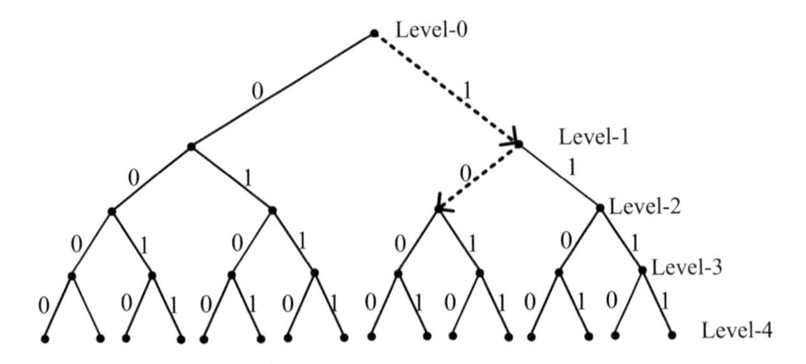

Fig. 3.5 Code-word formation using tree structure

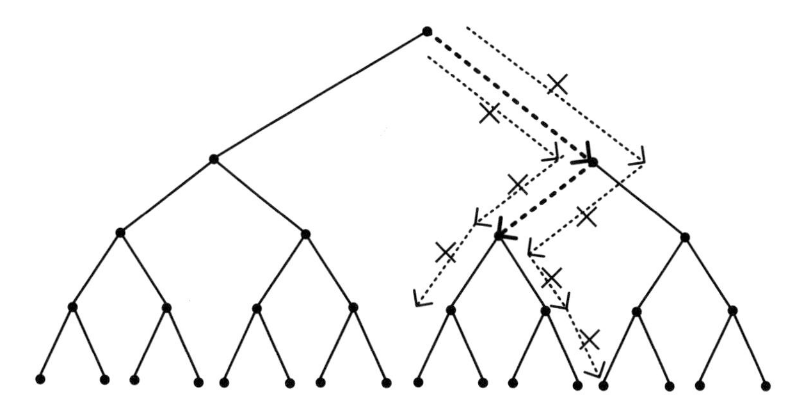

Fig. 3.6 Code-word formation using tree structure

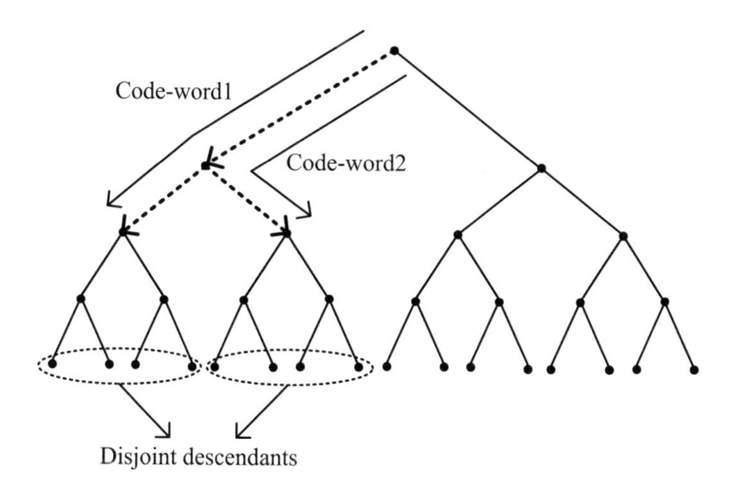

Fig. 3.7 Disjoint descendant nodes

While constructing the code-words using the code tree, assume that, you stop at a node at level l_i before going to the bottom most node, and collect the bits up to that node to construct the code-word. In this case, there are

$$2^{l_{max}-l_i}$$

descendant nodes which will never be employed for the construction of any other code-word due to the prefix condition. Considering Fig. 3.7, assume that there are N different code-words formed using code tree. In this case, there are N disjoint descendant sets, and each set includes

$$2^{l_{max}-l_i^c}$$

where l_i^c is the index of the level for the code-word c_i, number of descendants. On the other hand, the total number of descendant nodes at the bottom most layer equals to

$$2^{l_{max}}.$$

Assume that there are N code-words formed from the code tree. The total number of descendant nodes in the descendant sets equals to

$$\sum_{i=1}^{N} 2^{l_{max}-l_i^c}$$

which is less than or equal to the total number descendant nodes at the bottom most level, i.e.,

$$\sum_{i=1}^{N} 2^{l_{max}-l_i^c} \leq 2^{l_{max}}$$

which can be simplified as

$$\sum_{i=1}^{N} 2^{-l_i^c} \leq 1. \tag{3.57}$$

Inequality (3.57) is called the Kraft inequality. Now, let's state the Kraft inequality as a theorem, although we did already prove it.

Theorem 3.5 *The code-word lengths $l_1^c, l_2^c, \ldots$ of any prefix code or instantaneous code must satisfy*

$$\sum_{i=1}^{N} 2^{-l_i^c} \leq 1. \tag{3.58}$$

Conversely if a set of code-word lengths satisfy (3.58), *then there exist a prefix code with code-word lengths $l_1^c, l_2^c, \ldots$.*

Example 3.34 Let $\tilde{X}$ be a discrete random variable whose range set is $R_{\tilde{X}} = \{a, b, c, d, e\}$.

We want to design a prefix code whose code-word length are

$$l(c(a)) = 2, \quad l(c(b)) = 1, \quad l(c(c)) = 2, \quad l(c(d)) = 3, \quad l(c(e)) = 3.$$

Is it possible to design such a prefix code?

Solution 3.34 If the given code-word lengths satisfy the Kraft inequality

$$\sum_{i=1}^{N} 2^{-l_i^c} \leq 1$$

then it is possible to design a prefix code with the given code-word lengths. Substituting the given code-word length into (3.58), we get

$$2^{-2} + 2^{-1} + 2^{-2} + 2^{-3} + 2^{-3} \leq 1 \rightarrow 1.25 \leq 1$$

which is not correct. Thus, a prefix code with the given code-word lengths cannot be designed.

Example 3.35 Let $\tilde{X}$ be a discrete random variable whose range set is $R_{\tilde{X}} = \{a, b, c, d, e, f\}$. Design a prefix code for the symbols in $R_{\tilde{X}}$, and verify the Kraft inequality for the designed prefix code.

Solution 3.35 Using the code tree, we can choose the code-words according to our will as in Fig. 3.8.

As it is seen from Fig. 3.8, the code-word lengths are 4, 3, 2, 3, 3, 3. If Kraft inequality

$$\sum_{i=1}^{N} 2^{-l_i^c} \leq 1$$

is applied, we obtain

$$2^{-4} + 2^{-3} + 2^{-2} + 2^{-3} + 2^{-3} + 2^{-3} \leq 1$$

which is simplified as

$$\frac{15}{16} \leq 1.$$

Thus, Kraft inequality is satisfied for the designed code-word lengths.

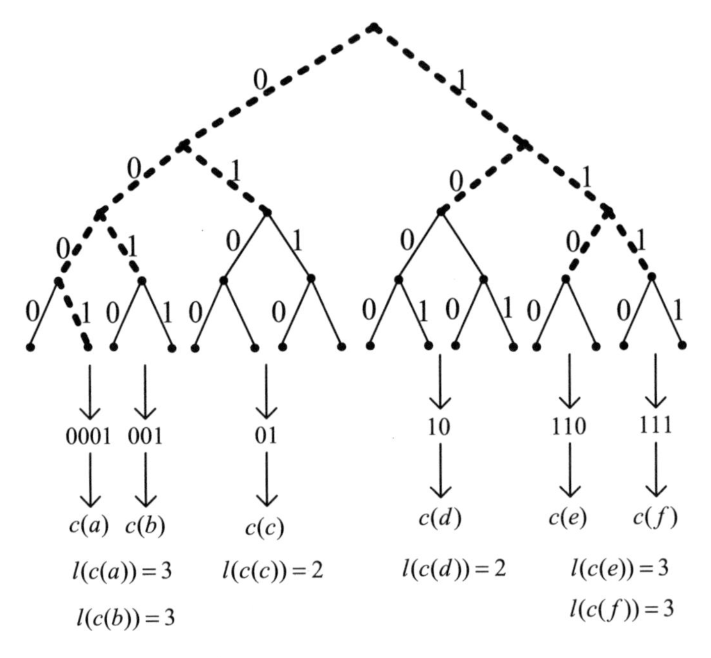

Fig. 3.8 Code-word formation using tree structure

3.4.2 *Optimal Codes*

We want to design a prefix code such that the expected length of the code-words is smaller than the expected length of the any other prefix code. In other words, we want to design the optimal prefix code. So, we should ask two questions for this purpose. First, what can be the minimum possible expected length of the code-words? Second, how can we construct optimal codes?

The Expected Code-Word Length of the Optimal Prefix Source Code:

Any prefix code satisfies the Kraft inequality

$$\sum_x 2^{l(c_x)} \leq 1. \tag{3.59}$$

The expected code-word length of any prefix code can be calculated using

$$L = \sum_x p(x) l(c_x). \tag{3.60}$$

Using (3.59) and (3.60), we can form the Lagrange multiplier as

$$\mathcal{L} = \sum_x p(x) l(c_x) + \lambda \left(\sum_x 2^{-l(c_x)} - 1 \right). \tag{3.61}$$

For the simplicity of the notation, let's use l_x for $l(c_x)$. Then, (3.61) becomes as

$$\mathcal{L} = \sum_x p(x) l_x + \lambda \left(\sum_x 2^{l_x} - 1 \right). \tag{3.62}$$

Taking the derivative of $\mathcal{L}$ with respect to l_x as

$$\frac{\partial \mathcal{L}}{\partial l_x} = p(x) - \lambda 2^{-l_x} \log_e 2$$

and equating to zero

$$p(x) - \lambda 2^{-l_x} \log_e 2 = 0$$

we obtain

$$2^{-l_x} = \frac{p(x)}{\lambda \log_e 2}. \tag{3.63}$$

Substituting the right hand side of (3.63) into

$$\sum_x 2^{-l_x} \leq 1$$

we get

$$\sum_x \frac{p(x)}{\lambda \log_e 2} \leq 1$$

from which, we obtain

$$\frac{1}{\log_e 2} \leq \lambda.$$

Choosing

$$\lambda = \frac{1}{\log_e 2}$$

and substituting it into (3.63), we get

$$2^{-l_x} = p(x)$$

from which, L_x is solved as

$$l_x^* = -\log p(x) \tag{3.64}$$

which is the optimum code-word length for symbol x. With the found optimum code-word lengths in (3.64), we can calculate the expected code-word length as

$$L^* = \sum_x p(x) l_x \rightarrow L^* = -\sum_x p(x) \log p(x) \rightarrow L^* = H(\tilde{X}). \tag{3.65}$$

We found in (3.65) that the expected code-word length of the optimum code equals to $H(\tilde{X})$, any other prefix code other than the optimum code has expected code-word length greater than $H(\tilde{X})$. Let's state this as a theorem and prove it.

Theorem 3.6 *Any prefix code other than the optimum prefix code has an expected code-word length greater than the expected code-word length of the optimum prefix code, in other words, if L_x is the expected code-word length of the any prefix code, than we have*

$$L_x \geq L_x^* = H(\tilde{X}) \tag{3.66}$$

where L_x^* denotes the expected code-word length of the optimum prefix code which is equal to entropy.

Proof 3.6 The expected code-word length of a prefix code is calculated as

$$L_x = \sum_x p(x) l_x$$

which can also be written as

$$L_x = -\sum_x p(x) \log 2^{-l_x}. \tag{3.67}$$

The expected optimum code-word length is found as

$$L^* = -\sum_x p(x) \log p(x). \tag{3.68}$$

Taking the difference of (3.67) and (3.68), we get

$$\begin{aligned} L_x - L^* &= -\sum_x p(x)\log 2^{-l_x} + \sum_x p(x)\log p(x) \\ &= -\sum_x p(x)\log \frac{2^{-l_x}}{p(x)} \end{aligned}$$

where employing the property

$$\log x \leq \frac{1}{\ln 2}(x-1)$$

we obtain

$$L_x - L^* \geq -\sum_x p(x)\frac{1}{\ln 2}\left(\frac{2^{-l_x}}{p(x)} - 1\right)$$

which is simplified as

$$L_x - L^* \geq \underbrace{\frac{1}{\ln 2}\left(\underbrace{\underbrace{\sum_x 2^{-l_x}}_{\substack{\leq 1 \\ \textit{Kraft Inequality}}} - \underbrace{\sum_x p(x)}_{=1}}_{\leq 0}\right)}_{\geq 0} \tag{3.69}$$

Thus, from (3.69), we obtain

$$L_x - L^* \geq 0 \rightarrow L_x \geq L^*.$$

Example 3.36 Let $\tilde{X}$ be a discrete random variable whose range set is $R_{\tilde{X}} = \{a, b, c, d\}$. The probability mass function of the random variable $\tilde{X}$ is given as

$$p(x=a) = \frac{1}{8} \quad p(x=b) = \frac{1}{8} \quad p(x=c) = \frac{2}{4} \quad p(x=c) = \frac{1}{4}.$$

Design prefix codes for the symbols in $R_{\tilde{X}}$, calculate the expected code-word lengths, and compare them to the expected code-word length of the optimal prefix code.

Solution 3.36 The entropy of the random variable $\tilde{X}$ can be calculated as

$$H(\tilde{X}) = -\sum_x p(x)\log p(x) \to H(\tilde{X}) = \frac{7}{4} \text{ bits/symbol.}$$

We can design a prefix code using the code tree as shown in Fig. 3.9.
Considering Fig. 3.9, we can calculate the expected code-word length of the designed code as

$$L_1 = \sum_x p(x)l_x \to L_x = \frac{1}{8}\times 3 + \frac{1}{8}\times 2 + \frac{2}{4}\times 2 + \frac{1}{4}\times 2 \to L_1 = \frac{17}{8}.$$

As we see

$$\frac{17}{8} = L_1 > H(\tilde{X}) = \frac{14}{8}.$$

We can design a better prefix code with the lower expected code-word length as shown in Fig. 3.10.
Considering Fig. 3.10, we can calculate the expected code-word length of the second designed code as

$$L_1 = \sum_x p(x)l_x \to L_x = \frac{1}{8}\times 2 + \frac{1}{8}\times 2 + \frac{2}{4}\times 2 + \frac{1}{4}\times 2 \to L_2 = 2.$$

As we see

$$(2 = L_2) > \left(H(\tilde{X}) = \frac{14}{8}\right).$$

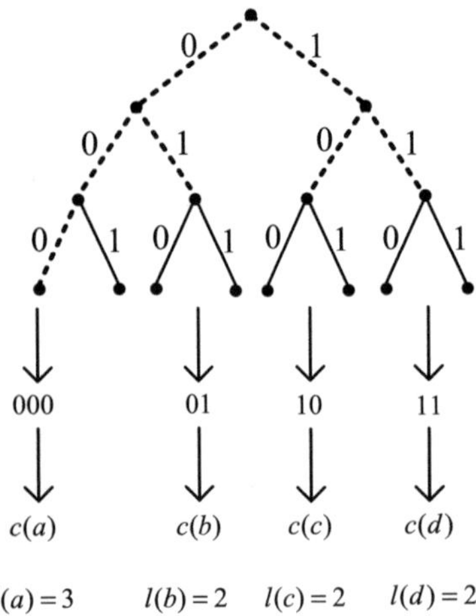

Fig. 3.9 Prefix code using the code tree

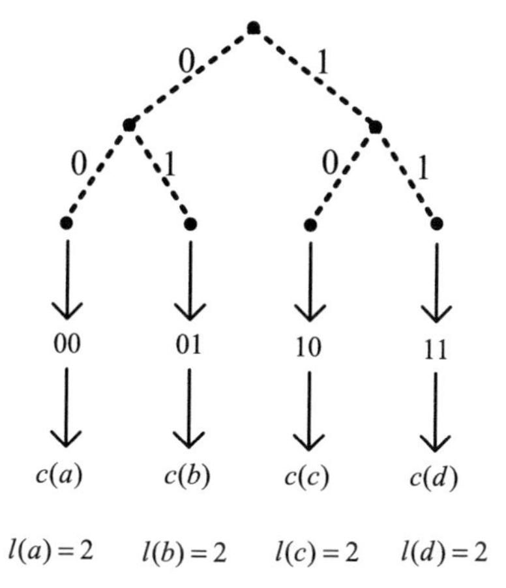

Fig. 3.10 Better prefix code using the code tree

If we compare the expected code-word lengths of the first and second code, we see that

$$\left(\frac{17}{8} = L_1\right) > (2 = L_2) > \left(H(\tilde{X}) = \frac{14}{8}\right).$$

Since the expected code-word length of the second code is closer to the entropy, we can say that the second source code is better than the first source code.

Design of Optimal Source Codes:

We have seen that the theoretical code-word lengths of the optimal prefix codes should be chosen according to

$$l(c(x)) = l_x = \log\frac{1}{p(x)}. \tag{3.70}$$

One of the optimal source codes proposed in the literature is the Shannon code. In Shannon codes, the code-word lengths are calculated as

$$l_x = \left\lceil \log\frac{1}{p(x)} \right\rceil \tag{3.71}$$

where $\lceil \cdot \rceil$ is the ceil operator which rounds the number to the closest larger integer. Shannon codes are nothing but a class of prefix codes.

Example 3.37 Show that (3.71) satisfies the Kraft inequality.

Solution 3.37 Kraft inequality implies that

$$\sum_x 2^{-l_x} \leq 1 \tag{3.72}$$

Equation (3.71) means that

$$l_x = \log \frac{1}{p(x)} + \epsilon \tag{3.73}$$

where $0 < \epsilon < 1$. Substituting (3.73) into (3.72), we get

$$\sum_x 2^{-\left(\log \frac{1}{p(x)} + \epsilon\right)} \leq 1 \tag{3.74}$$

which can be written as

$$2^{-\epsilon} \sum_x 2^{-\log \frac{1}{p(x)}} \leq 1 \tag{3.75}$$

which can be simplified as

$$2^{-\epsilon} \underbrace{\sum_x p(x)}_{=1} \leq 1 \rightarrow 2^{-\epsilon} \leq 1 \rightarrow 1 \leq 2^{\epsilon}$$

which is a correct inequality. Thus, (3.71) satisfies the Kraft inequality, i.e., Shannon codes are prefix codes.

Bounds on Average Code-Word Length of Shannon Code

Equation (3.71) which is restated as

$$l_x = \left\lceil \log \frac{1}{p(x)} \right\rceil$$

implies that

$$l_x = \log \frac{1}{p(x)} + \epsilon, \quad 0 \leq \epsilon < 1$$

which means that

$$\log\frac{1}{p(x)} \leq l_x < \log\frac{1}{p(x)} + 1. \tag{3.76}$$

Multiplying all sides of (3.76) by $p(x)$ and summing over x, we get

$$\underbrace{\sum_x p(x)\log\frac{1}{p(x)}}_{H(\tilde{X})} \leq \underbrace{\sum_x p(x)\, l_x}_{L_x^s} < \underbrace{\sum_x p(x)\left(\log\frac{1}{p(x)} + 1\right)}_{H(\tilde{X})+1}$$

which can be written as

$$H(\tilde{X}) \leq L_x^s < H(\tilde{X}) + 1 \tag{3.77}$$

where L_x^s is the expected code-word length of the Shannon code. We can restate the obtained result as a theorem.

Theorem 3.7 *The code-word lengths of the Shannon codes are decided according to*

$$l_x = \left\lceil \log\frac{1}{p(x)} \right\rceil.$$

It can be shown that the expected code-word length of the Shannon code satisfies

$$H(\tilde{X}) \leq L_x^s < H(\tilde{X}) + 1 \tag{3.78}$$

where $H(\tilde{X})$ is the entropy of the random variable $\tilde{X}$.

Example 3.38 For the discrete random variable $\tilde{X}$, the range set $R_{\tilde{X}}$ and the probability mass function $p(x)$ are given as

$$R_{\tilde{X}} = \{a, b\}$$
$$p(x = a) = \tfrac{4}{5}, \quad p(x = b) = \tfrac{1}{5}.$$

Determine the Shannon code-word lengths, and verify the inequality

$$H(\tilde{X}) \leq L_x^* < H(\tilde{X}) + 1. \tag{3.79}$$

Solution 3.38 The entropy of the random variable $\tilde{X}$ can be calculated as

$$H(\tilde{X}) = -\sum_x p(x)\log p(x) \rightarrow H(\tilde{X}) = -\left(\tfrac{4}{5}\log\tfrac{4}{5} + \tfrac{1}{5}\log\tfrac{1}{5}\right) \rightarrow H(\tilde{X}) = 0.7219\,\text{bits/symbol}$$

The code-word lengths for the symbols a and b are found using

$$l_x = \left\lceil \log\frac{1}{p(x)} \right\rceil$$

as

$$l_a = \left\lceil \log\frac{1}{p(a)} \right\rceil \rightarrow l_a = \left\lceil \log\frac{5}{4} \right\rceil \rightarrow l_a = 0.32 \rightarrow l_a = 1\,bit.$$

$$l_b = \left\lceil \log\frac{1}{p(b)} \right\rceil \rightarrow l_b = \left\lceil \log\frac{5}{1} \right\rceil \rightarrow l_b = 2.32 \rightarrow l_b = 3\,\text{bits}.$$

The average code-word length can be calculated using

$$L_x^s = \sum_x p(x)l_x \rightarrow L_x = \frac{4}{5}\times 1 + \frac{1}{5}\times 3 \rightarrow L_x = 1.4\,\text{bits}.$$

If we check the inequality

$$H(\tilde{X}) \leq L_x^s < H(\tilde{X}) + 1 \tag{3.80}$$

for the calculated values, we obtain

$$0.7219 \leq 1.4 < 0.7219 + 1.$$

Thus, we verified the inequality in (3.80).

3.4.3 *Source Coding for Real Number Sequences*

In this section, we will consider finding code-words for sequences of symbols instead of finding a code-word for each single symbol. Let's denote the sequence of IID random variables $\tilde{X}_1, \tilde{X}_2, \ldots, \tilde{X}_N \sim p(x)$ by $\tilde{Y}$, i.e.,

$$\tilde{Y} = \left(\tilde{X}_1, \tilde{X}_2, \ldots, \tilde{X}_N\right).$$

Assume that Shannon code is constructed for $\tilde{Y}$. In this case, the expected code-word length for $\tilde{Y}$ satisfies

$$H(\tilde{Y}) \leq L^s_{x_1^N} < H(\tilde{Y}) + 1$$

in which substituting $\tilde{Y} = (\tilde{X}_1, \tilde{X}_2, \ldots, \tilde{X}_N)$, we get

$$H(\tilde{X}_1, \tilde{X}_2, \ldots, \tilde{X}_N) \leq L^s_{x_1^N} < H(\tilde{X}_1, \tilde{X}_2, \ldots, \tilde{X}_N) + 1 \tag{3.81}$$

Since the random variables $\tilde{X}_1, \tilde{X}_2, \ldots, \tilde{X}_N$ are independent of each other, the joint entropy $H(\tilde{X}_1, \tilde{X}_2, \ldots, \tilde{X}_N)$ in (3.81) can be written as

$$H(\tilde{X}_1, \tilde{X}_2, \ldots, \tilde{X}_N) = \sum_{i=1}^{N} H(\tilde{X}_i). \tag{3.82}$$

Substituting (3.82) into (3.81), we obtain

$$\sum_{i=1}^{N} H(\tilde{X}_i) \leq L^s_{x_1^N} < \sum_{i=1}^{N} H(\tilde{X}_i) + 1. \tag{3.83}$$

Using the generic random variable $\tilde{X} \sim p(x)$, we can write

$$H(\tilde{X}_i) = H(\tilde{X}). \tag{3.84}$$

Using (3.84) in (3.83), we get

$$\sum_{i=1}^{N} H(\tilde{X}) \leq L^s_{x_1^N} < \sum_{i=1}^{N} H(\tilde{X}) + 1 \tag{3.85}$$

which leads to

$$NH(\tilde{X}) \leq L^s_{x_1^N} < NH(\tilde{X}) + 1. \tag{3.86}$$

Equation (3.86) implies that the real number sequence $x_1, x_2, \ldots, x_N$ generated by the random variable sequence $\tilde{X}_1, \tilde{X}_2, \ldots, \tilde{X}_N$ can be represented by

$$\lceil NH(\tilde{X}) \rceil$$

number of bits. This means that each symbol in the sequence is represented by

$$\frac{\lceil NH(\tilde{X}) \rceil}{N}$$

number of bits on average. Besides, from (3.86), we can write that

$$H(\tilde{X}) \leq \frac{L^s_{x_1^N}}{N} < H(\tilde{X}) + \frac{1}{N} \tag{3.87}$$

where

$$\frac{L^s_{x_1^N}}{N}$$

is the number of bits used to represent a symbol. We see that the bounds in the inequality

$$H(\tilde{X}) \leq \frac{L^s_{x_1^N}}{N} < H(\tilde{X}) + \frac{1}{N} \tag{3.88}$$

is tighter than the bounds in

$$H(\tilde{X}) \leq L^s < H(\tilde{X}) + 1 \tag{3.89}$$

which was obtained when source coding is performed for each symbol separately. If we compare the inequalities (3.88) and (3.89), we see that designing source codes for sequence of symbols directly is a more efficient method, i.e., we use less number of bits per-symbol on average. Let's illustrate this concept by an example.

Example 3.39 For the discrete random variable $\tilde{X}$, the range set $R_{\tilde{X}}$ and the probability mass function $p(x)$ are given as

$$R_{\tilde{X}} = \{a, b\}$$

$$p(x = a) = \frac{4}{5} \quad p(x = b) = \frac{1}{5}.$$

(a) Determine Shannon code-word lengths for symbol sequences consisting of two symbols, and find the average code-word length used for sequences.
(b) Repeat part-a for symbol sequences consisting of three symbols.

Solution 3.39 (a) We consider symbol sequence sequences x_1^N where $N = 2$. The possible symbol sequences can be formed as

$$aa,\ ab,\ ba,\ bb.$$

We can consider each sequence as a symbol and find the probability of each sequence using $p(x_1, x_2, \ldots, x_N) = p(x_1)p(x_2)\cdots p(x_N)$ as

$$p(aa) = p(a)p(a) \rightarrow p(aa) = \frac{16}{25}$$
$$p(ab) = p(a)p(b) \rightarrow p(ab) = \frac{4}{25}$$
$$p(ba) = p(b)p(a) \rightarrow p(ba) = \frac{4}{25}$$
$$p(bb) = p(b)p(b) \rightarrow p(bb) = \frac{1}{25}.$$

The Shannon code-word length for each sequence can be calculated using

$$l_{x_1^N} = \left\lceil \log \frac{1}{p(x_1^N)} \right\rceil$$

as

$$l_{aa} = \left\lceil \log \frac{1}{p(aa)} \right\rceil \rightarrow l_{aa} = \left\lceil \log \frac{25}{16} \right\rceil \rightarrow l_{aa} = 1$$
$$l_{ab} = \left\lceil \log \frac{1}{p(ab)} \right\rceil \rightarrow l_{ab} = \left\lceil \log \frac{25}{4} \right\rceil \rightarrow l_{ab} = 3$$
$$l_{ba} = \left\lceil \log \frac{1}{p(ba)} \right\rceil \rightarrow l_{ba} = \left\lceil \log \frac{25}{4} \right\rceil \rightarrow l_{ba} = 3$$
$$l_{bb} = \left\lceil \log \frac{1}{p(bb)} \right\rceil \rightarrow l_{bb} = \left\lceil \log \frac{25}{1} \right\rceil \rightarrow l_{bb} = 5.$$

The average code-word length can be calculated using

$$L_{x_1^N}^s = \sum_{x_1^N} p(x_1^N) l_{x_1^N}$$

leading to

$$L^s_{x_1^2} = \frac{16}{25} \times 1 + \frac{4}{25} \times 3 + \frac{4}{25} \times 3 + \frac{1}{25} \times 5 \rightarrow L_{x_1^2} = 1.8\,\text{bits}$$

which means that we use 1.8 bits to represent a symbol sequence consisting of two symbols. This implies that a single symbol in the sequence is represented by $1.8/2 = 0.9$ bits.
In Example 3.38, we found that $H(\tilde{X}) = 0.72$ bits/symbol. If we check the bound

$$H(\tilde{X}) \leq \frac{L^s_{x_1^N}}{N} < H(\tilde{X}) + \frac{1}{N}$$

for $N = 2$, we see that

$$0.72 \leq \frac{1.8}{2} < 0.72 + \frac{1}{2}$$

which is simplified as

$$0.72 \leq 0.9 < 1.22$$

which is a tighter bound than

$$0.72 \leq 1.4 < 1.72$$

obtained in Example 3.38.
(b) Let's now consider symbol sequences consisting of three symbols each. The possible symbol sequences can be written as

$$aaa,\ aab,\ aba,\ baa,\ abb,\ bab,\ bba,\ bbb. \tag{3.90}$$

Let's denote the symbol sequences in (3.90) by the symbols $s_1,\ s_2, s_3,\ s_4,\ s_5,\ s_6,\ s_7,$ and s_8 respectively, i.e., s_1 indicates *aaa*, and s_8 indicates *bbb*.

The probabilities of these sequences can be calculated as

$$\begin{aligned}
p(s_1) &= p(a)p(a)p(a) \rightarrow p(s_1) = \frac{64}{125} \\
p(s_2) &= p(a)p(a)p(b) \rightarrow p(s_2) = \frac{16}{125} \\
p(s_3) &= p(a)p(b)p(a) \rightarrow p(s_3) = \frac{16}{125} \\
p(s_4) &= p(b)p(a)p(a) \rightarrow p(s_4) = \frac{16}{125} \\
p(s_5) &= p(a)p(b)p(b) \rightarrow p(s_5) = \frac{4}{125} \\
p(s_6) &= p(b)p(a)p(b) \rightarrow p(s_6) = \frac{4}{125} \\
p(s_7) &= p(b)p(b)p(a) \rightarrow p(s_7) = \frac{16}{25} \\
p(s_8) &= p(b)p(b)p(b) \rightarrow p(s_8) = \frac{4}{25}.
\end{aligned} \tag{3.91}$$

Considering the calculated probabilities in (3.91), we can calculate the Shannon code-word lengths using

$$l_{s_i} = \left\lceil \log \frac{1}{p(s_i)} \right\rceil$$

as

$$\begin{aligned}
l_{s_1} &= \left\lceil \log \frac{125}{64} \right\rceil \rightarrow l_{s_1} = 1 \\
l_{s_2} &= \left\lceil \log \frac{125}{16} \right\rceil \rightarrow l_{s_2} = 3 \\
l_{s_3} &= \left\lceil \log \frac{125}{16} \right\rceil \rightarrow l_{s_3} = 3 \\
l_{s_4} &= \left\lceil \log \frac{125}{16} \right\rceil \rightarrow l_{s_4} = 3 \\
l_{s_5} &= \left\lceil \log \frac{125}{4} \right\rceil \rightarrow l_{s_5} = 5 \\
l_{s_6} &= \left\lceil \log \frac{125}{4} \right\rceil \rightarrow l_{s_6} = 5 \\
l_{s_7} &= \left\lceil \log \frac{125}{16} \right\rceil \rightarrow l_{s_7} = 3 \\
l_{s_8} &= \left\lceil \log \frac{125}{4} \right\rceil \rightarrow l_{s_8} = 5.
\end{aligned}$$

Using the calculated code-word lengths, we can find the expected code-word length using

$$L^s_{x_1^N} = \sum_i p(s_i) l_{s_i} \tag{3.92}$$

as

$$L^s_{x_1^3} = \frac{64}{125} \times 1 + \frac{16}{125} \times 3 \times 4 + \frac{4}{125} \times 5 \times 3 \rightarrow L_{x_1^N} \approx 2.53 \text{ bits}$$

which means that we use 2.538 bits to represent a symbol sequence consisting of three symbols. This implies that a single symbol in the sequence is represented by $2.538/3 \approx 0.84$ bits.

In Example 3.38, we found that $H(\tilde{X}) = 0.72$ bits/symbol. And in part-a we found that if we represent sequences consisting of two symbols, we use 0.9 bits for each symbol. However, in this part we found that if we represent sequences consisting of three symbols, we use 0.84 bits for each symbol. We see that as the length of represented sequences increase, less number of bits are used to represent each symbol on average.

If we check the bound

$$H(\tilde{X}) \leq \frac{L^s_{x_1^N}}{N} < H(\tilde{X}) + \frac{1}{N}$$

for $N = 3$, we see that

$$0.72 \leq \frac{2.538}{3} < 0.72 + \frac{1}{3}$$

which is simplified as

$$0.72 \leq 0.84 < 1.05$$

which is a tighter bound than the bound

$$0.72 \leq 0.9 < 1.22$$

found in part-a.

Until now in our examples, we only considered code-word lengths, however, we did not make a discussion how to find those code-words with the calculated code-word lengths. In fact, this is an engineering work. Designing good source codes is still an ongoing research, and better source codes are introduced in time.

We will now introduce one of the optimal source codes designed by Huffman. This source code is called Huffman code. Huffman codes are constructed using the probability mass function of the information sources, in other words, it depends on

the statistics of the information sources. Although, Huffman codes are very efficient codes, their source dependency makes them impractical for systems employing different kind of information sources. For instance, assume that we have a Huffman source encoder for speech files recorded in English, then this source encoder system would be useless for speech files recorded in Turkish, in other words, the encoder is not a generic type encoder with great flexibility.
Let's briefly visit the Huffman source codes.

3.4.4 *Huffman Codes*

Huffman codes are practical optimal codes in the sense that their expected code-word length is the smallest considering other prefix codes. In fact, Huffman codes are prefix type codes. Huffman codes are constructed by combining small probabilities until we get '1'. The Huffman code construction is illustrated in the next example.

Example 3.40 The range set of a discrete random variable $\tilde{X}$ is given as

$$R_{\tilde{X}} = \{a, b, c, d, e\}$$

for which the probability mass function is defined as

$$p(a) = 0.2 \quad p(b) = 0.4 \quad p(c) = 0.15 \quad p(d) = 0.01 \quad p(e) = 0.24.$$

(a) Find the expected code-word length of the Shannon code.
(b) Design a Huffman code and calculate the expected code-word length and compare it to the expected code-word length of the Shannon code.

Solution 3.40

(a) The expected code-word length of the Shannon code is calculated using

$$L_x^s = \sum_x p(x) \left\lceil \log \frac{1}{p(x)} \right\rceil \tag{3.93}$$

leading to

$$L_x^s = p(a) \left\lceil \log \frac{1}{p(a)} \right\rceil + p(b) \left\lceil \log \frac{1}{p(b)} \right\rceil + p(c) \left\lceil \log \frac{1}{p(c)} \right\rceil + p(d) \left\lceil \log \frac{1}{p(d)} \right\rceil + p(e) \left\lceil \log \frac{1}{p(e)} \right\rceil$$

which can be written as

$$L_x^s = 0.2\left\lceil \log\frac{1}{0.2}\right\rceil + 0.4\left\lceil \log\frac{1}{0.4}\right\rceil + 0.15\left\lceil \log\frac{1}{0.15}\right\rceil + 0.01\left\lceil \log\frac{1}{0.01}\right\rceil + 0.24\left\lceil \log\frac{1}{0.24}\right\rceil$$

which is simplified as

$$L_x^s = 0.2 \times 3 + 0.4 \times 2 + 0.15 \times 3 + 0.01 \times 7 + 0.24 \times 3$$

resulting in

$$L_x^s = 2.64\,\text{bits}.$$

On the other hand, the theoretical minimum value of the expected code-word length is

$$L_x^* = p(a)\log\frac{1}{p(a)} + p(b)\log\frac{1}{p(b)} + p(c)\log\frac{1}{p(c)} + p(d)\log\frac{1}{p(d)} + p(e)\log\frac{1}{p(e)}$$

resulting in

$$L_x^* = 1.9643\,\text{bits}.$$

We see that expected value of Shannon code-words is far away from the theoretical minimum possible value, i.e., $2.64 > 1.9643$.
(b) In Huffman encoding, we first sort the probabilities, and their corresponding symbols in decreasing order as shown in Fig. 3.11.
Then, we sum the smallest two probabilities as illustrated in Fig. 3.12.
Next, we add the previous summation result to the next lowest probability as shown in Fig. 3.13.
This procedure is repeated till the largest probability, and at the end we get the probabilities in Fig. 3.14.
In the last step, we label the horizontal lines by ‘0’, and sloped lines by ‘1’ as shown in Fig. 3.15 and to find the code-words for the symbols we move from the tip of the triangle to the symbol along the lines and collect the bits. The formation of the code-word for symbol a is illustrated in Fig. 3.16.

b	0.4
e	0.24
a	0.2
c	0.15
d	0.01

Fig. 3.11 Sorted probabilities

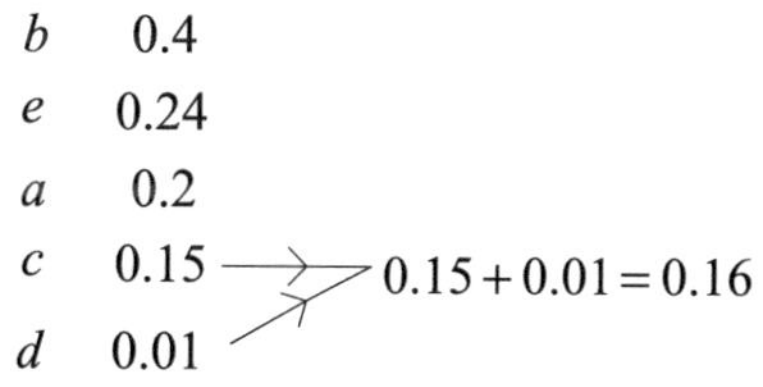

b 0.4
e 0.24
a 0.2
c 0.15
d 0.01
$0.15 + 0.01 = 0.16$

Fig. 3.12 Probability summation

b 0.4
e 0.24
a 0.2
c 0.15
d 0.01
0.16
$0.16 + 0.2 = 0.36$

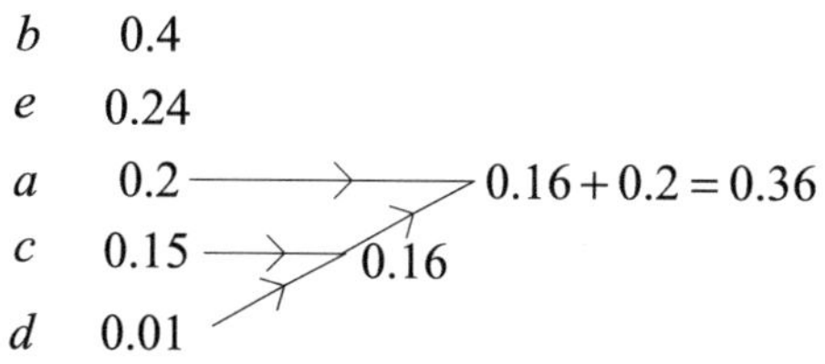

Fig. 3.13 Probability summation

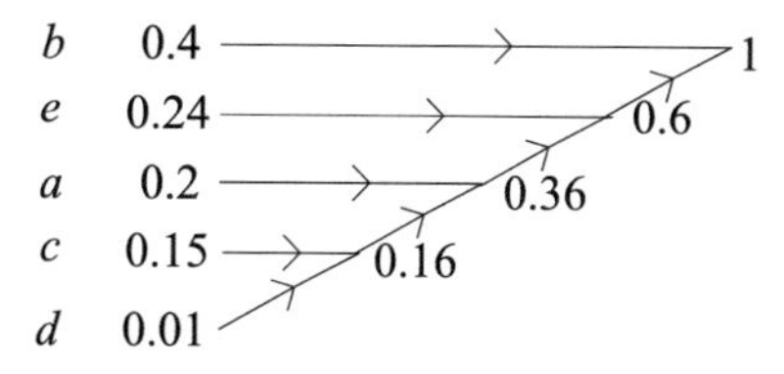

Fig. 3.14 Probability summation

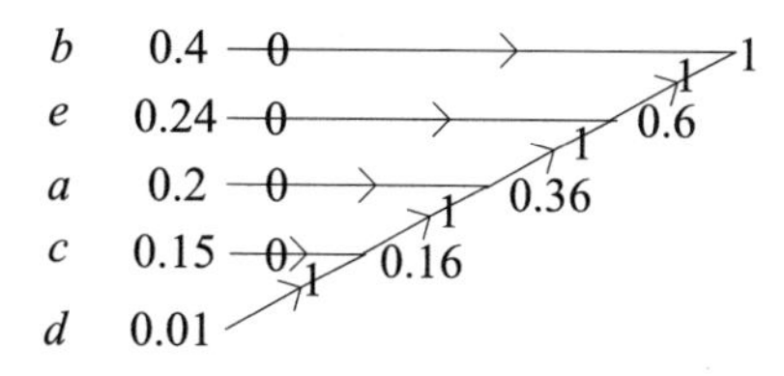

Fig. 3.15 Labeling the lines

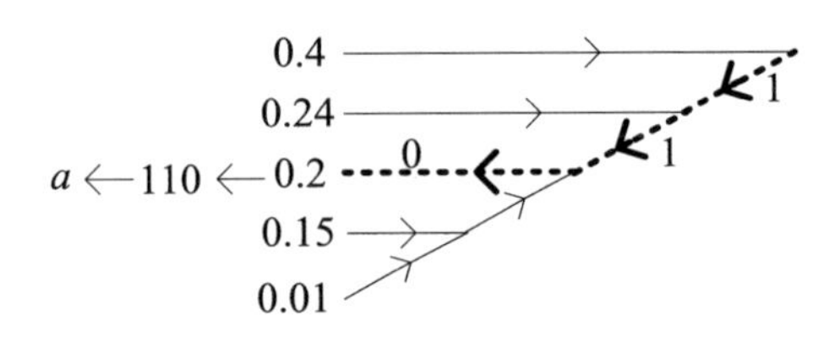

Fig. 3.16 Formation of prefix codes

Following a similar approach for the other symbols, we find the code-words as shown in (3.94)

$$\begin{aligned} b &\to 0 \\ e &\to 10 \\ a &\to 110 \\ c &\to 1110 \\ d &\to 1111. \end{aligned} \tag{3.94}$$

From (3.94), we can write the code-word lengths for the symbols a, b, c, d, e as

$$l_b = 1, \quad l_e = 2, \quad l_a = 3, \quad l_c = 4, \quad l_d = 4. \tag{3.95}$$

The expected code-word length of the Huffman code can be calculated as

$$L_x = 0.4 \times 1 + 0.24 \times 2 + 0.2 \times 3 + 0.15 \times 4 + 0.01 \times 4 \to L_x = 2.12\,\text{bits}.$$

We see that for this example, the Huffman code has better expected code-word length than the expected code-word length of the Shannon codes, since, Shannon code has expected code-word length 2.46 which is greater than 2.12.

Huffman codes can also be designed for sequence of symbols. In this case, we should first calculate the probabilities of the sequences, and follow the same procedure explained in the example. It is expected that if we encode the sequences instead of symbols alone, we would obtain more efficient source codes. However, as the length of the sequences increases, it may not be possible to calculate the probability of sequences due to very large number of sequences.

Exercise For the discrete random variable $\tilde{X}$, the range set $R_{\tilde{X}}$ and the probability mass function $p(x)$ are given as

$$R_{\tilde{X}} = \{a, b\}$$
$$p(x = a) = \tfrac{4}{5} \quad p(x = b) = \tfrac{1}{5}.$$

(a) Write all symbol sequences consisting of two symbols, and design a Huffman code for the two-symbol sequences. Calculate the expected code-word length.

(b) Repeat part-a for symbol sequences consisting of three symbols.
(c) Repeat part-a for the discrete random variable $\tilde{X}$ given as

$$R_{\tilde{X}} = \{a,\, b,\, c,\, d,\, e\}$$

for which the probability mass function is defined as

$$p(a) = 0.2 \quad p(b) = 0.4 \quad p(c) = 0.15 \quad p(d) = 0.01 \quad p(e) = 0.24.$$

Problems

(1) What does x_1^6 mean?
(2) We have IID random variable sequence

$$\tilde{X}_1,\, \tilde{X}_2, \ldots, \tilde{X}_N \sim p(x)$$

and range set $R_{\tilde{X}} = \{a,\, b,\, c\}$. The probability mass function $p(x)$ is defined on the range set as

$$p(x = a) = \frac{1}{8} \quad p(x = b) = \frac{6}{8} \quad p(x = c) = \frac{1}{8}$$

and the generic random variable having the same probability mass function is indicated as

$$\tilde{X} \sim p(x).$$

(a) For $N = 8$, find some typical sequences. How many typical sequences are available in the typical set? Find also the number of non-typical sequences. Calculate the probability of typical sequence you found.

(3) What is the difference between T_ϵ^4 and T_x^4?
(4) Let $\tilde{X}_1, \tilde{X}_2, \ldots, \tilde{X}_N \sim p(x)$, and $\tilde{X} \sim p(x)$. The range set for $\tilde{X}$ is defined as

$$R_{\tilde{X}} = \{a,\, b,\, e\}.$$

Write some sample sequences for x_1^N where $N = 3$ and $N = 7$.
(5) Let $\tilde{X}_1, \tilde{X}_2, \ldots, \tilde{X}_N \sim p(x)$, and $\tilde{X} \sim p(x)$. The range set for $\tilde{X}$ and probability mass function $p(x)$ is defined as

$$R_{\tilde{X}} = \{a,\, b, c\}$$
$$p(x = a) = \tfrac{1}{16} \quad p(x = b) = \tfrac{3}{16} \quad p(x = c) = \tfrac{4}{16}.$$

Find the number of typical sequences for $N = 30$, and $N = 80$.

(6) For the random variable $\tilde{X}$ the range set and probability mass function are given as

$$R_{\tilde{X}} = \{a, b, c\}$$
$$p(x = a) = \tfrac{2}{4} \quad p(x = b) = \tfrac{1}{4} \quad p(x = c) = \tfrac{1}{4}.$$

(a) Let $\tilde{X}_1^N \sim p(x)$. Find a strongly typical sequence x_1^N for $N = 20$ and verify the inequality

$$2^{-N(H(\tilde{X})+\epsilon)} \leq p(x_1, x_2, \ldots, x_N) \leq 2^{-N(H(\tilde{X})-\epsilon)}$$

for your sequence.

(b) Find the total number of strongly typical sequences, and weakly typical sequences.

(7) Let $\tilde{X}_1, \tilde{X}_2, \ldots, \tilde{X}_N \sim p(x)$, and $\tilde{X} \sim p(x)$. The range set for $\tilde{X}$ and probability mass function $p(x)$ is defined as

$$R_{\tilde{X}} = \{a, b, c\}$$
$$p(x = a) = \tfrac{3}{8} \quad p(x = b) = \tfrac{2}{8} \quad p(x = c) = \tfrac{3}{8}.$$

Find the number of typical and non-typical sequences for $N = 24$.

(8) Let $\tilde{X}_1, \tilde{X}_2, \ldots, \tilde{X}_N \sim p(x)$, and $\tilde{X} \sim p(x)$. The range set for $\tilde{X}$ and probability mass function $p(x)$ is defined as

$$R_{\tilde{X}} = \{a, b, c\}$$
$$p(x = a) = \tfrac{1}{8} \quad p(x = b) = \tfrac{2}{8} \quad p(x = c) = \tfrac{5}{8}.$$

Find the number of typical and non-typical sequences for $N = 24$. How many bits are required to represent a typical sequence?

(9) For the discrete random variable $\tilde{X}$, the range set $R_{\tilde{X}}$, and the probability mass function $p(x)$ are given as

$$R_{\tilde{X}} = \{a, b, c, d\},$$
$$p(x = a) = \tfrac{1}{4} \quad p(x = b) = \tfrac{1}{8} \quad p(x = c) = \tfrac{1}{8} \quad p(x = d) = \tfrac{1}{2}.$$

The binary code-words for the symbols in the range set are defined as

$$c(a) = 01, \quad c(b) = 10, \quad c(c) = 110, \quad c(d) = 101.$$

Find the expected value, i.e., probabilistic average, of the code-words

(10) Let $\tilde{X}$ be a discrete random variable whose range set is $R_{\tilde{X}} = \{a, b, c, d, e\}$. We want to design a prefix code whose code-word length are

$$l(c(a)) = 4, \quad l(c(b)) = 1, \quad l(c(c)) = 3, \quad l(c(d)) = 1, \quad l(c(e)) = 2.$$

Is it possible to design such a prefix code?

(11) For the discrete random variable $\tilde{X}$, the range set $R_{\tilde{X}}$ and the probability mass function $p(x)$ are given as

$$R_{\tilde{X}} = \{a, b\}$$
$$p(x = a) = \tfrac{2}{7} \quad p(x = b) = \tfrac{5}{7}.$$

Determine the Shannon code-word lengths, and verify the inequality

$$H(\tilde{X}) \leq L_x^* < H(\tilde{X}) + 1.$$

(12) For the discrete random variable $\tilde{X}$, the range set $R_{\tilde{X}}$, and the probability mass function $p(x)$ are given as

$$R_{\tilde{X}} = \{a, b\}$$
$$p(x = a) = \tfrac{1}{8} \quad p(x = b) = \tfrac{7}{8}.$$

(a) Determine Shannon code-word lengths for symbol sequences consisting of two symbols, and find the average code-word length used for sequences.
(b) Repeat part-a for symbol sequences consisting of three symbols.

(13) For the discrete random variable $\tilde{X}$, the range set $R_{\tilde{X}}$ and the probability mass function $p(x)$ are given as

$$R_{\tilde{X}} = \{a, b\}$$
$$p(x = a) = \tfrac{1}{4} \quad p(x = b) = \tfrac{3}{4}.$$

(a) Write all symbol sequences consisting of two symbols, and design a Huffman code for the two-symbol sequences. Calculate the expected code-word length.
(b) Repeat part-a for symbol sequences consisting of three symbols.

Chapter 4
Channel Coding Theorem

In this chapter, we will discuss the channel coding theorem in details. Channel coding theorem can be accepted as a milestone in electronic communication field. For this reason, it is very critical for a communication engineer to comprehend the channel coding theorem very well. In many information theory books, or in many lecture notes delivered in classes about information theory, channel coding theorem is very briefly summarized, for this reason, many readers fail to comprehend the details behind the theorem.

Channel coding theorem is a mathematically involved concept. Considering this issue, we provided many simple examples before discussing the channel coding theorem. To understand the philosophy of this theorem very well, we will first provide some preparation material on the subject, then proceed with the proof of the theorem.

4.1 Discrete Memoryless Channel

Definition Discrete memoryless channel:
A discrete memoryless channel has an input random variable and output random variable. An example discrete channel is shown in Fig. 4.1.
The input random variable has the probability mass function $p(x)$ and the output random variable has the probability mass function $p(y)$. The channel transition probabilities are nothing but the conditional probabilities $p(y|x)$. The range sets of the input and output random variables are indicated by $R_{\tilde{X}}$ and $R_{\tilde{Y}}$. For the example channel in Fig. 4.1, $R_{\tilde{X}} = \{x_1, x_2\}$ and $R_{\tilde{Y}} = \{y_1, y_2\}$.

The discrete memoryless channel can also be represented as a black-box as in Fig. 4.2.

O. Gazi, *Information Theory for Electrical Engineers*, Signals and Communication Technology, https://doi.org/10.1007/978-981-10-8432-4_4

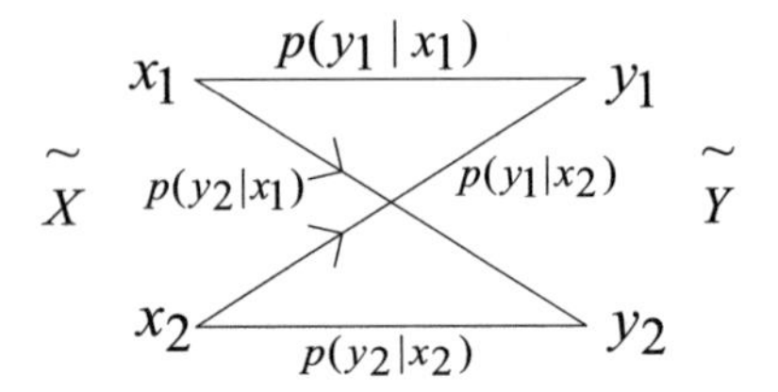

Fig. 4.1 A discrete memoryless channel

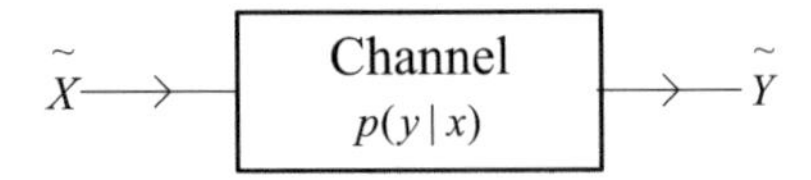

Fig. 4.2 Black-box representation of discrete memoryless channel

A discrete memoryless channel is mathematically characterized by

$$(R_{\tilde{X}}, p(y|x), R_{\tilde{Y}}) \tag{4.1}$$

Definition Extended discrete memoryless channel:
If a discrete memoryless channel is repeatedly used N times, this is indicated as

$$\left(R_{\tilde{X}^N}, p\left(y^N|x^N\right), R_{\tilde{Y}^N}\right) \tag{4.2}$$

where $y^N = y_1 y_2 \ldots y_N$, and $x^N = x_1 x_2 \ldots x_N$ are real number sequences. An example extended discrete memoryless channel is shown in Fig. 4.3.
Note: Memoryless condition implies that

$$p\left(y_k|x^k, y^{k-1}\right) = p(y_k|x_k) \tag{4.3}$$

where $x^k = x_1 x_2 \ldots x_k$ and $y^{k-1} = y_1 y_2 \ldots y_{k-1}$.
And we will use the term 'discrete channel' for the place of 'discrete memoryless channel' unless otherwise indicated.

4.2 Communication System

A typical communication system is depicted in Fig. 4.4

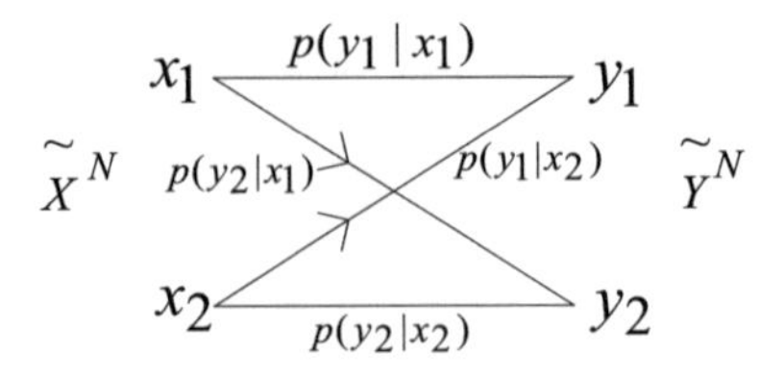

Fig. 4.3 Extended discrete memoryless channel

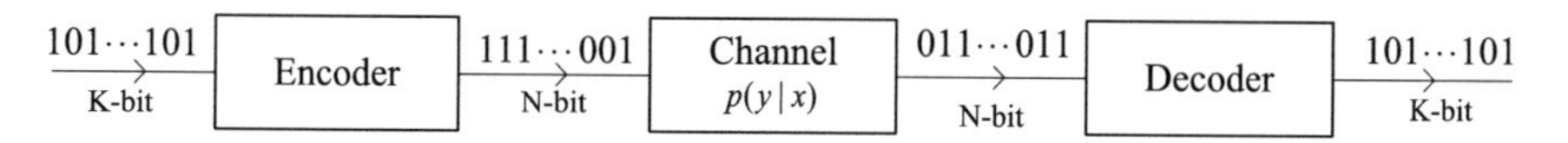

Fig. 4.4 A typical communication system

where K-bit sequence is the information message sent thought the communication system. Using K bits , it is possible to generate

$$L = 2^K \tag{4.4}$$

different information sequences, and to distinguish these sequences from each other, we can assign an index value starting from 1 to each sequence. This is illustrated in Table 4.1 for $K = 8$, and $L = 2^K$.

The communication system in Fig. 4.4 can be represented using the random variable sequences as shown in Fig. 4.5 where $\tilde{M}$ is the message index random variable with range set

$$R_{\tilde{M}} = \{1, 2, \ldots, L = 2^K\} \tag{4.5}$$

m is the message index representing a K-bit information sequence. $\tilde{X}^N(m)$ is the random variable sequence for the message index m, i.e.,

Table 4.1 Assigning indexes to sequences

Sequence	Index
00000000	1
00000001	2
00000010	3
⋮	⋮
⋮	⋮
11111111	L

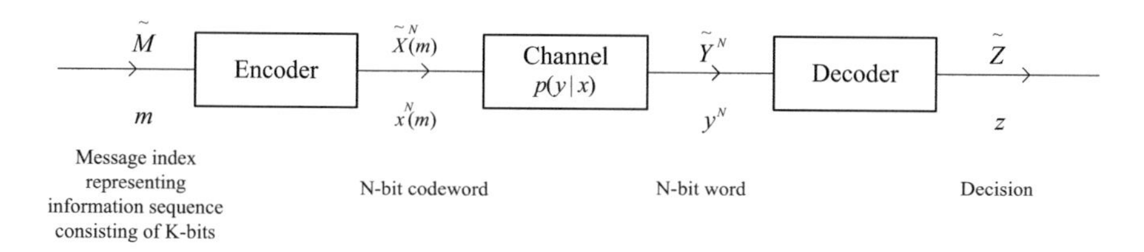

Fig. 4.5 A typical communication system in detail

$$\tilde{X}^N(m) = \tilde{X}_{m1}\tilde{X}_{m2}\ldots\tilde{X}_{mN} \tag{4.6}$$

$x^N(m)$ generated by $\tilde{X}^N(m)$ is the code-word for message index m and it consists of N bits, i.e.,

$$x^N(m) = x_{m1}x_{m_2}\ldots x_{mN} \tag{4.7}$$

$\tilde{X}^N(m)$ and $x^N(m)$ can be considered as functions. Their input is the message index m, their outputs are the random variable sequences, and real number sequences respectively.

$\tilde{Y}^N$ is the random variable sequence at the output of the channel, i.e.,

$$\tilde{Y}^N = \tilde{Y}_1\tilde{Y}_2\ldots\tilde{Y}^N \tag{4.8}$$

y^N generated by $\tilde{Y}^N$ is the bit vector at the output of the channel consisting of N bits, i.e.,

$$y^N = y_1y_2\ldots y_N \tag{4.9}$$

and z is the estimate of the decoder about the transmitted message index.

At the encoder output we have N bits, and $N > K$. The ratio

$$R = \frac{K}{N} \tag{4.10}$$

is called the rate of the encoder, or the rate of the code. The number of information sequences L can also be expressed as

$$L = 2^{NR}. \tag{4.11}$$

The decoder takes $\tilde{Y}^N$ as input and produces $\tilde{Z}$, this operation can be expressed by

$$\tilde{Z} = d\left(\tilde{Y}^N\right) \tag{4.12}$$

where $d(\cdot)$ is the decoding function. The random variable $\tilde{Z}$ has the range set

$$R_{\tilde{Z}} = \left\{1, 2, \ldots, L = 2^K\right\} \tag{4.13}$$

which is the same range set of $\tilde{M}$.

4.2.1 Probability of Error

Definition Conditional probability of error is given as

$$\beta_i = Prob(\tilde{Z} \neq i|\tilde{M} = i) \tag{4.14}$$

which can also be expressed as

$$\beta_i = Prob(d(\tilde{Y}^N) \neq i|\tilde{X}^N(i) = x^N(i)) \tag{4.15}$$

which can be calculated using

$$\beta_i = \sum_{\substack{y^N \\ d(y^N) \neq i}} p(y^N|x^N(i)) \tag{4.16}$$

Example 4.1 For the communication system given in Fig. 4.6, $K = 2$, $N = 3$. The data-words, their corresponding code-words, the index values and the decoding function $d(\cdot)$ are described in Table 4.2.

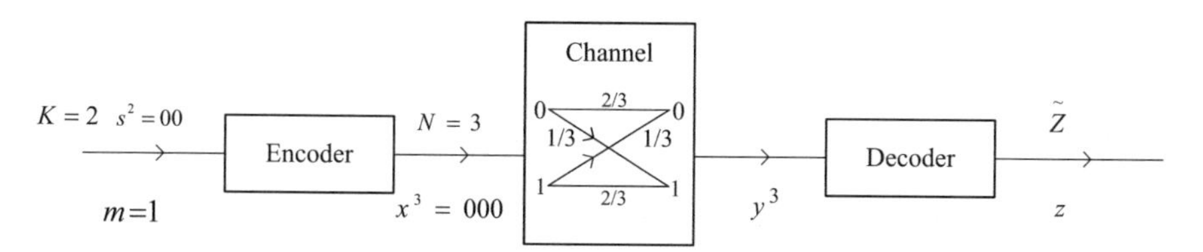

Fig. 4.6 Communication system for Example 4.1

Table 4.2 Data-words, code-words, indexes, and decoding function

Data-word K = 2 s^2	Index, m	Code-word N = 3 x^3(m)	Decoding function $d(y^3)$
00	1	000	$d(000) = 1$ $d(001) = 1$
01	2	011	$d(010) = 2$ $d(011) = 2$
10	3	101	$d(100) = 2$ $d(101) = 2$
11	4	111	$d(011) = 2$ $d(111) = 2$

Assume that all zero data word is transmitted as shown in Fig. 4.6, i.e., $s^2 = 00$. Calculate the conditional probability of error for the transmitted corresponding code-word.

Solution 4.1 For the transmitted data-word $s^2 = 00$, the generated code-word is $x^3 = 000$. We can calculate the conditional probability of error for the transmitted code word using

$$\beta_i = \sum_{\substack{\tilde{Y}^N \\ d(y^N) \neq i}} p(y^N | x^N(i))$$

as

$$\beta_1 = \sum_{\substack{\tilde{Y}^N \\ d(y^3) \neq 1}} p(y^3 | x^3(1)) \tag{4.17}$$

Using Table 4.2, we can calculate (4.17) as

$$\beta_0 = p(010|000) + p(011|000) + p(100|000) + p(101|000) + p(011|000) + p(111|000)$$

in which employing $p(y_3 y_2 y_1 | x_3 x_2 x_1) = p(y_3|x_3)p(y_2|x_2)p(y_1|x_1)$ for each term in the summation using the channel described in Fig. 4.6, we get

$$\beta_1 = \left(\frac{2}{3}\right)^2 \left(\frac{1}{3}\right) + \left(\frac{2}{3}\right)\left(\frac{1}{3}\right)^2 + \left(\frac{2}{3}\right)^2 \left(\frac{1}{3}\right) + \left(\frac{2}{3}\right)\left(\frac{1}{3}\right)^2 + \left(\frac{2}{3}\right)\left(\frac{1}{3}\right)^2 + \left(\frac{1}{3}\right)^3$$

resulting in

$$\beta_1 = 0.5556.$$

Definition Maximal probability of error:
The maximal probability of error for a channel code described as (N, K) is calculated using

$$\beta^* = \max_{i \in \{1,2,\ldots,L\}} \beta_i. \tag{4.18}$$

Example 4.2 For Example 4.1, we calculated $\beta_1 = 0.5556$. If we calculate the other conditional probabilities, we find them as

$$\beta_2 = 0.5185 \, \beta_3 = 0.5556 \, \beta_4 = 0.5556$$

then, β^* is found as

$$\beta^* = \max_{i \in \{1,2,\ldots,4\}} \beta_i \rightarrow \beta^* = 0.5556.$$

Definition Arithmetic average of probability of error P_a for an (N, K) code is defined as

$$P_a = \frac{1}{L} \sum_{i=1}^{L} \beta_i \leq \beta^*. \tag{4.19}$$

Example 4.3 For Example 4.2, the arithmetic average of probability of error can be calculated as

$$P_a = \frac{(\beta_1 + \beta_2 + \beta_3 + \beta_4)}{4} \rightarrow P_a = 0.5463.$$

4.2.2 *Rate Achievability*

Definition The rate of an (N, K) is defined as

$$R = \frac{K}{N}. \tag{4.20}$$

Considering that $L = 2^K$, the rate of the code can also be expressed as

$$R = \frac{\log L}{N}. \tag{4.21}$$

Example 4.4 The rate of a code is given as $R = 0.7$. What do you infer from the given information?

Solution $R = 0.7$ implies that every transmitted code-bit carries 0.7 bits of data and 0.3 bits of redundancy which is used to decrease the probability of error during transmission. That means that 100 code-bits carry 70 bits of data and 30 bits of redundancy.

Definition The rate R is said to be achievable, if we can find an (N, K) code such that the maximal probability of the code goes to zero as the length of the code-words goes to infinity, i.e., R is achievable if

$$\beta^* \to 0 \quad \text{as} \quad N \to \infty. \tag{4.22}$$

Definition: Capacity can also be interpreted as the supremum of all achievable code rates.

Example 4.5 In a paper, it is written that the code rate $R = 0.5$ is achievable. Does it mean that if I encode 3 bits of information vector as 6 bits of code-word I can recover 3 bits of information at the decoder side?

Solution It does not mean that you can recover your data at the decoder side if you encode your 3-bit data with a code of rate $R = 0.5$. The meaning of achievability for rate $R = 0.5$ is explained in Fig. 4.7.

The concept of achievability of the code rate R is illustrated in Fig. 4.7 where it is seen that as the values of K and N increases the ratio $R = K/N$ stays constant, and for sufficiently large lengths of K and N if we can have $\beta^* \to 0$, then we say that the rate $R = K/N$ is achievable.

Example 4.6 Let's illustrate the meaning of the capacity as follows

$$\begin{gathered} R = 0.3 \text{ is achievable} \\ R = 0.4 \text{ is achievable} \\ R = 0.5 \text{ is achievable} \\ \vdots \\ R = 0.8 \text{ is achievable} \\ R > 0.8 \text{ is NOT achievable} \end{gathered}$$

Dataword Length	Codeword Length	Rate	Maximal Probability of Error	Rate Achievability
$K = 2$ 01	$N = 4$ 0011 $\longrightarrow$	$R = 2/4$ $R = 0.5$	$\beta^* \not\to 0$	Not Achievable
$K = 3$ 010	$N = 6$ 001100 $\longrightarrow$	$R = 3/6$ $R = 0.5$	$\beta^* \not\to 0$	Not Achievable
$K = 4$ 0101	$N = 4$ 00110011 $\longrightarrow$	$R = 4/8$ $R = 0.5$	$\beta^* \not\to 0$	Not Achievable
⋮	⋮	⋮	⋮	⋮
$K = 100$ 01⋯10	$N = 200$ 0011⋯1100 $\longrightarrow$	$R = 100/200$ $R = 0.5$	$\beta^* \longrightarrow 0$	**ACHIEVABLE**

Fig. 4.7 Illustration of rate achievability

Then, we can say that capacity equals to $C = 0.8$.

Example 4.7 Capacity is defined as the maximum amount of mutual information between input and output a discrete memoryless channel, but, now it is defined as the maximum achievable rate of a channel code. Shouldn't they be the same? I am a bit confused from these two definitions. Can you a bit more explain the capacity concept in details with an example?

Solution 4.7 Let's consider the communication system depicted in Fig. 4.8 where the channel employed is a binary erasure channel with erasure probability $p = 1/4$. So, the capacity of the binary erasure channel can be calculated as

$$C = 1 - p \rightarrow C = 1 - \frac{1}{4} \rightarrow C = \frac{3}{4}.$$

The capacity $C = 3/4$ implies that if we transmit 400 bits through the channel repeatedly at most 300 of them can be reliably received in average. However, this is the theoretical limit. To get 300 information bits at the output of the channel, or at the output of the communication system, we add the encoder and decoders units to the communication system as shown in Fig. 4.8. The rate of the encoder in Fig. 4.8 is $R = 3/4$. Assume that the rate is achievable, i.e., the transmitted 300 data bits are recoverable at the output of the decoder. This means that, we are transmitting at the capacity limit.

Assume that we are using a lower rate code, such as $R = 3/8$, and assume that this rate is achievable. This means that for every 300 data-bits encoder produces 800 code-bits, and these code-bits are transmitted through binary erasure channel. At the output of the channel we have 800 code-bits, and using these 800 bits at the decoder we can recover 300 transmitted data bits.

We can increase the rate of the code used in the system, such as $R = 3/7$, $R = 3/6$, but our highest rate cannot be larger than $R = 3/4$, which is decided from the capacity of the binary erasure channel, for reliable transmission.

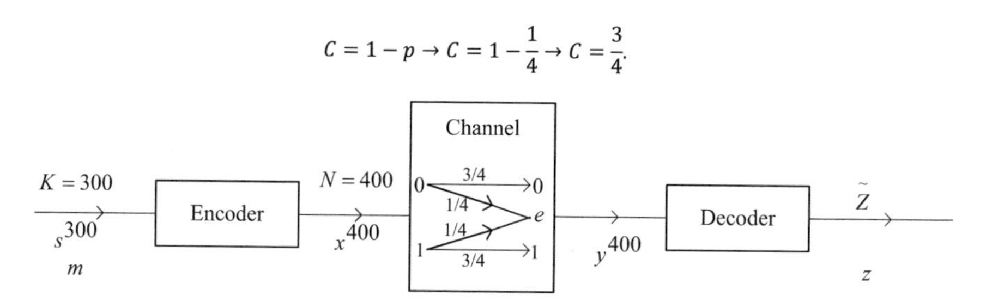

Fig. 4.8 Communication system for Example 4.7

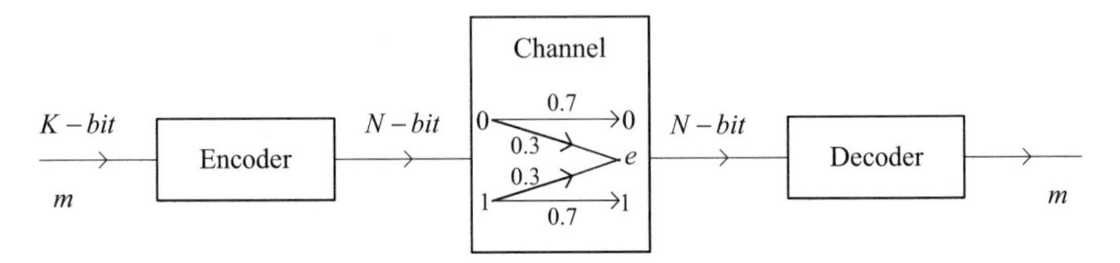

Fig. 4.9 Communication system for Example 4.8

Example 4.8 For the communication system given in Fig. 4.9, it is reported that the code rate $R = 0.82$ is achievable. Decide whether the given information in correct or not.

Solution 4.8 The erasure probability of the binary erasure channel is $\alpha = 0.3$, and the capacity of the binary erasure channel is $C = 1 - \alpha \rightarrow C = 0.7$. The rate R is achievable if it satisfies $R < C$. However, for the given R and C values $R < C$ leads to $0.82 < 0.7$ which is incorrect inequality.

Thus, we can say that the provided information in the question is not correct, i.e., $R = 0.82$ cannot be achieved for the communication system given in Fig. 4.9.

4.3 Jointly Typical Sequences

The sequences x^N and y^N are jointly typical sequences if

1. x^N *is a typical sequence* w.r.t. $p(x^N)$, *i.e.*, $\left|-\frac{1}{N}\log p(x^N) - H(\tilde{X})\right| < \epsilon$
2. y^N *is a typical sequence* w.r.t. $p(y^N)$, *i.e.*, $\left|-\frac{1}{N}\log p(y^N) - H(\tilde{Y})\right| < \epsilon$
3. (x^N, y^N) *is jointly typical* w.r.t. $p(x^N, y^N)$, *i.e.*, $\left|-\frac{1}{N}\log p(x^N, y^N) - H(\tilde{X}, \tilde{Y})\right| < \epsilon$

And we have

$$p(x^N, y^N) = \prod_{i=1}^{N} p(x_i, y_i).$$

The criteria for the sequences x^N and y^N to be jointly typical can also be expressed as

1. x^N *is a typical sequence* w.r.t., $p(x^N)$

$$i.e., \ 2^{-N\left(H(\tilde{X}) + \epsilon\right)} \leq p(x^n) \leq 2^{-N\left(H(\tilde{X}) - \epsilon\right)}$$

2. y^N *is a typical sequence w.r.t.,* $p(y^N)$

$$i.e., 2^{-N\left(H(\tilde{Y})+\epsilon\right)} \leq p(y^n) \leq 2^{-N\left(H(\tilde{Y})-\epsilon\right)}$$

3. (x^N, y^N) *is a typical pair w.r.t.,* $p(x^N, y^N)$

$$i.e., 2^{-N\left(H(\tilde{X},\tilde{Y})+\epsilon\right)} \leq p(x^n, y^n) \leq 2^{-N\left(H(\tilde{X},\tilde{Y})-\epsilon\right)}$$

4.3.1 Jointly Typical Set

The set T_ϵ^N consisting of jointly typical sequences (x^N, y^N) is called jointly typical set.
Note: x^N indicates the sequence $x_1x_2...x_N$.

4.3.2 Strongly and Weakly Jointly Typical Sequences

In jointly typical pair (x^N, y^N), if (x_i, y_i) appears $N \times p(x_i, y_i)$ times, then the jointly typical sequence (x^N, y^N) is called strongly jointly typical sequence, otherwise, it is called weakly jointly typical sequence.

Let's solve some examples to understand the concept of jointly typical sequences better.

Example 4.9 Let's consider the binary symmetric channel shown in Fig. 4.10.

At the channel input, we have a discrete random variable $\tilde{X}$ with the range set $R_{\tilde{X}} = \{0, 1\}$, and at the channel output, we have a discrete random variable $\tilde{Y}$ with the range set $R_{\tilde{Y}} = \{0, 1\}$. Assume that the channel transition probability p equals to $1/4$. And the probability mass function of $\tilde{X}$ is given as

$$p(x=0) = \frac{1}{2}\, p(x=1) = \frac{1}{2}.$$

(a) Find the probability mass function of $\tilde{Y}$, i.e., $p(y) = ?$
(b) Find the joint probability mass function between $\tilde{X}$ and $\tilde{Y}$, i.e., $p(x, y) = ?$
(c) Comment on the use of binary symmetric channel in Fig. 4.10.

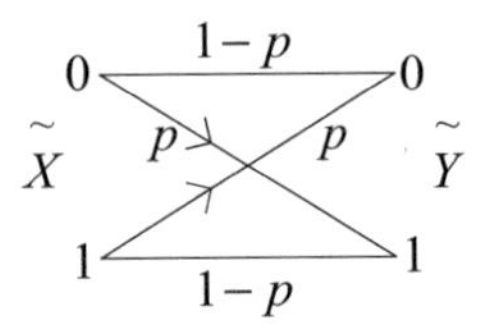

Fig. 4.10 Binary symmetric channel for Example 4.9

Solution 4.9 The probability mass function $p(y)$ can be calculated using

$$p(y) = \sum_x p(x, y) \rightarrow p(y) = \sum_x p(y|x)p(x)$$

from which we obtain

$$p(y = 0) = \sum_x p(y = 0|x)p(x) \rightarrow$$

$$p(y = 0) = \underbrace{p(y = 0|x = 0)}_{1-p}\underbrace{p(x = 0)}_{\frac{1}{2}} + \underbrace{p(y = 0|x = 1)}_{p}\underbrace{p(x = 1)}_{\frac{1}{2}} \rightarrow$$

$$p(y = 0) = \frac{1}{2}$$

and for $p(y = 1)$, we have

$$p(y = 1) = \sum_x p(y = 1|x)p(x) \rightarrow$$

$$p(y = 1) = \underbrace{p(y = 1|x = 0)}_{p}\underbrace{p(x = 0)}_{\frac{1}{2}} + \underbrace{p(y = 1|x = 1)}_{1-p}\underbrace{p(x = 1)}_{\frac{1}{2}} \rightarrow$$

$$p(y = 1) = \frac{1}{2}.$$

(b) The joint probability mass function $p(x,y)$ can be calculated using

$$p(x, y) = p(y|x)p(x)$$

leading to

$$p(x = 0, y = 0) = p(y = 0|x = 0)p(x = 0) \rightarrow p(x = 0, y = 0) = \frac{3}{8}$$

$$p(x = 0, y = 1) = p(y = 1|x = 0)p(x = 0) \rightarrow p(x = 0, y = 1) = \frac{1}{8}$$

$$p(x = 1, y = 0) = p(y = 0|x = 1)p(x = 1) \rightarrow p(x = 1, y = 0) = \frac{1}{8}$$

$$p(x = 1, y = 1) = p(y = 1|x = 1)p(x = 1) \rightarrow p(x = 1, y = 1) = \frac{3}{8}$$

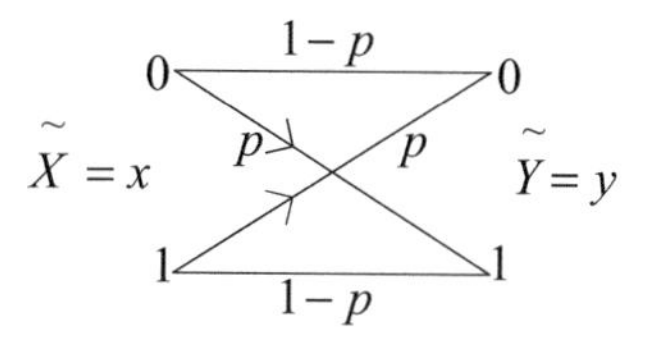

Fig. 4.11 The binary symmetric channel

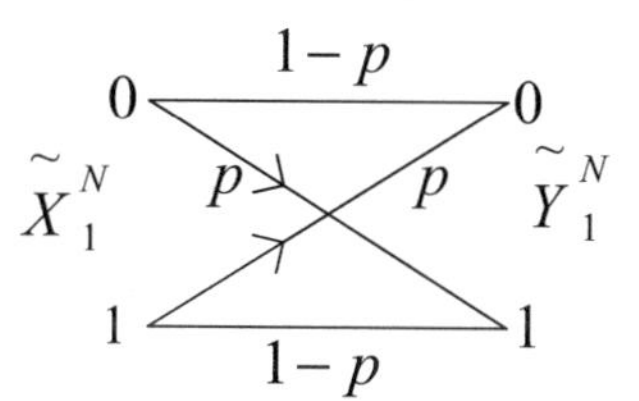

Fig. 4.12 Extended binary symmetric channel for Example 4.10

(c) Let's now comment on the use of channel. The binary symmetric channel shown in Fig. 4.10 implies that the random variable $\tilde{Y}$ generates a value x which can be equal to either 0 or 1 as depicted in Fig. 4.11. If the generated value is $x = 0$, it is more probable to have $y = 0$ at the channel output, since $p(x = 0, y = 0) = 3/8$ is greater than $p(x = 0, y = 1) = 1/8$.

Example 4.10 Let's consider the extended binary symmetric channel shown in Fig. 4.12 where $\tilde{X}_1^N \sim p(x)$, $\tilde{Y}_1^N \sim p(y)$ and we have the generic random variables $\tilde{X} \sim p(x), \tilde{Y} \sim p(y)$ with range sets $R_{\tilde{X}} = \{0, 1\}$ and $R_{\tilde{Y}} = \{0, 1\}$. Interpret the use of channel in Fig. 4.12.

Solution 4.10 The communication channel in Fig. 4.12 implies the system in Fig. 4.13. That is, the binary symmetric channel is used N times. All the input random variables at the input of the channels have the same probability mass function and range set, and all the output random variables at the output of the channels have the same probability mass function and range set.

Example 4.11 Consider the extended binary symmetric channel shown in Fig. 4.14 where $\tilde{X}_1^N \sim p(x)$, $\tilde{Y}_1^N \sim p(y)$ and we have the generic random variables $\tilde{X} \sim p(x), \tilde{Y} \sim p(y$ with range sets $R_{\tilde{X}} = \{0, 1\}$, $R_{\tilde{Y}} = \{0, 1\}$, and $p(x = 0) = 1/2$, $p(x = 0) = 1/2$.

Let the sequence length be N = 8. The transition probability is p = 1/4. Assume that at the input of the channel, we have the typical sequence $x^8 = 10110100$ generated by the random variable sequence $\tilde{X}^8$. Find a sequence y^8 at the output of the channel such that x^8 and y^8 are jointly typical.

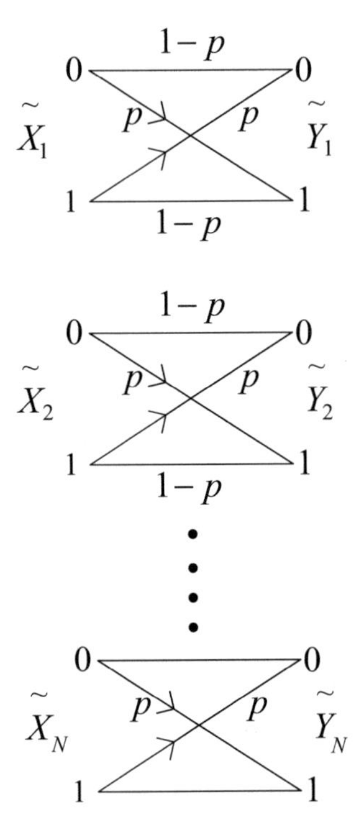

Fig. 4.13 Extended binary symmetric channel in details

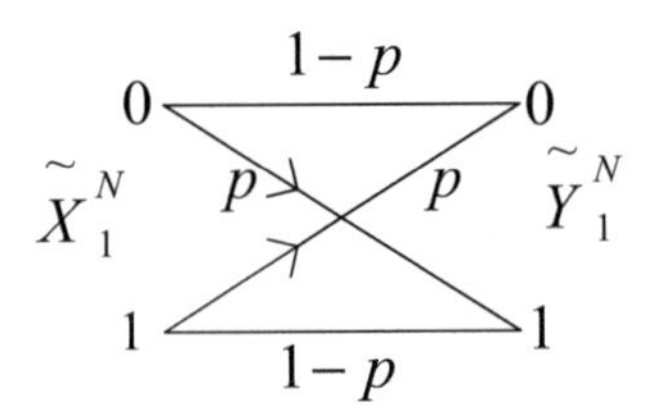

Fig. 4.14 Extended binary symmetric channel for Example 4.11

Solution 4.11 In our previous Example 4.9, we found that $p(y=0)=1/2$ and $p(y=1)=1/2$. We can calculate the entropies $H(\tilde{X})$ and $H(\tilde{Y})$ as

$$H(\tilde{X}) = H(\tilde{X}) = 1\,\text{bits/symbol}. \tag{4.23}$$

First, let's check whether the given sequence is typical or not. If it is a typical sequence, then it should satisfy

$$2^{-N\left(H(\tilde{X})+\epsilon\right)} \leq p\left(x^N\right) \leq 2^{-N\left(H(\tilde{X})-\epsilon\right)}. \quad (4.24)$$

Since the random variables $\tilde{X}_1^8$ are IID, we can write

$$p\left(x^8\right) = p(x_1)p(x_2)\ldots p(x_8) \quad (4.25)$$

leading to

$$p\left(x^8\right) = \left(\frac{1}{2}\right)^6 \left(\frac{1}{2}\right)^2 \rightarrow p\left(x^8\right) = \left(\frac{1}{2}\right)^8. \quad (4.26)$$

Let's choose $\epsilon = 0.01$, substituting (4.26) and (4.23) into (4.24), we get

$$2^{-8(1+0.01)} \leq \left(\frac{1}{2}\right)^8 \leq 2^{-8(1-0.01)} \quad (4.27)$$

leading to

$$0.0037 \leq 0.0039 \leq 0.0041$$

which is correct, then, the given sequence is a typical sequence.
Using the joint probability mass function calculated in Example 4.9, the joint entropy $H(\tilde{X}, \tilde{Y})$ between $\tilde{X}$ and $\tilde{Y}$ can be calculated using

$$H\left(\tilde{X}, \tilde{Y}\right) = -\sum_{x,y} p(x,y) \log p(x,y)$$

as

$$H\left(\tilde{X}, \tilde{Y}\right) = 1.8113 \text{ bits}/symbol\text{–pair}.$$

The jointly typical condition implies that

$$2^{-N\left(H(\tilde{X},\tilde{Y})+\epsilon\right)} \leq p\left(x^N, y^N\right) \leq 2^{-N\left(H(\tilde{X},\tilde{Y})-\epsilon\right)}$$

in which substituting $N = 8, \epsilon = 0.01$ and $H\left(\tilde{X}, \tilde{Y}\right) = 1.8113$ we get

$$2^{-8(1.8113+0.01)} \leq p\left(x^8, y^8\right) \leq 2^{-8(1.8113-0.01)}$$

leading to

$$4.11 \times 10^{-5} \leq p\left(x^8, y^8\right) \leq 4.59 \times 10^{-5}. \quad (4.28)$$

In the question, x^8 is given as $x^8 = 10110100$. Let $y^8 = 01110100$, i.e., we flipped the first two bits at the beginning of x^8, and using $p(x^8, y^8) = p(x_1, y_1)p(x_2, y_2)\ldots p(x_8, y_8)$ we can calculate $p(x^8, y^8)$ as

$$p(x^8, y^8) = (p(x = 1, y = 1))^6 \times (p(x = 0, y = 0))^2$$

leading to

$$p(x^8, y^8) = \left(\frac{3}{8}\right)^6 \left(\frac{1}{8}\right)^2 \rightarrow p(x^8, y^8) = 4.35 \times 10^{-5}$$

which satisfies (4.28). Thus, $x^8 = 10110100$ and $y^8 = 01110100$ are jointly typical sequences.

We indicated that to find y^8 that is jointly typical with $x^8 = 10110100$, we flipped the first 2 bits in $x^8 = 10110100$. In fact, if the flipping probabilities of the channel in Fig. 4.14 is inspected we see that they are equal to $1/4$ for both '0' and '1'. In our sequence $x^8 = 10110100$, we have four '0's and four '1's. If we flip $1/4$ of the total number of zeros, we flip $4/4 = 1$ zero only, and if we flip $1/4$ of the total number of ones, we flip $4/4 = 1$ one only, and in our sequence $x^8 = 10110100$ the first two bits are '1' and '0'. That is why flipping the first two bits of $x^8 = 10110100$ leaded us to a sequence that is jointly typical with $x^8 = 10110100$.

In fact, by flipping a single '0' and a single '1' in $x^8 = 10110100$ we can obtain other different sequences which are jointly typical with x^8. This means that for the given sequence $x^8 = 10110111$ we can find $\binom{4}{1}\binom{4}{1}$ different y^8 jointly typical with x^8.

Example 4.12 Consider the extended binary symmetric channel shown in Fig. 4.15 where $\tilde{X}_1^{20} \sim p(x)$, $\tilde{Y}_1^{20} \sim p(y)$ and we have the generic random variables $\tilde{X} \sim p(x), \tilde{Y} \sim p(y)$ with range sets $R_{\tilde{X}} = \{0, 1\}$, $R_{\tilde{Y}} = \{0, 1\}$, and $p(x = 0) = 1/2$, $p(x = 1) = 1/2$.

Assume that we have a bit-vector consisting of 20 bits at the input of the channel, i.e., x^{20}. How do you find a bit-vector at the output of the channel which is jointly typical with x^{20}?

Solution 4.12 In Fig. 4.15, we see that the input and output random variables have uniform distribution and the flipping probability of the binary symmetric channel is $1/4$. This implies that as the length of the input sequence goes to infinity, it is certain that $1/4$ of the total number of zeros and $1/4$ of the total number of ones in

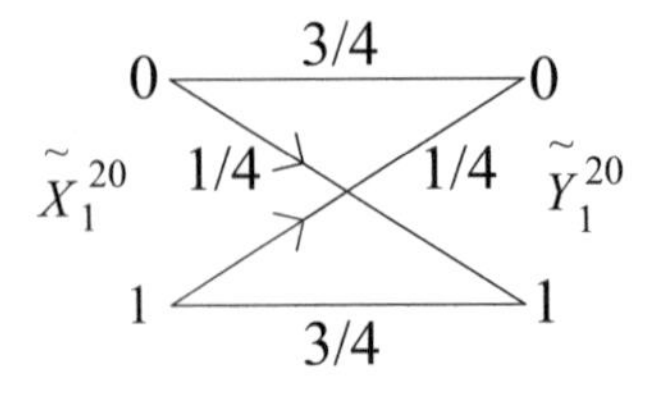

Fig. 4.15 Extended binary symmetric channel for Example 4.12

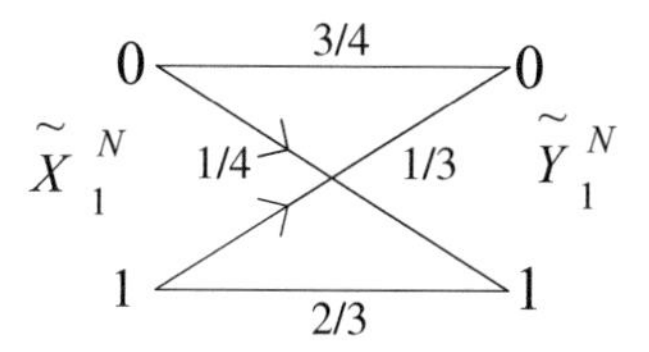

Fig. 4.16 Extended binary channel for Example 4.13

the sequence x^N will be flipped. To find a sequence that it jointly typical with x^{20}, it is sufficient to flip $1/4$ of the zeros and $1/4$ of ones in x^{20}, and denote it by y^{20}.

Example 4.13 Consider the extended binary channel shown in Fig. 4.16 where $\tilde{X}_1^N \sim p(x)$, $\tilde{Y}_1^N \sim p(y)$ and we have the generic random variables $\tilde{X} \sim p(x), \tilde{Y} \sim p(y)$ with range sets $R_{\tilde{X}} = \{0, 1\}$, $R_{\tilde{Y}} = \{0, 1\}$, and $p(x=0) = 3/4, p(x=1) = 1/4$. At the input of the channel, assume that we have a typical sequence x^N generated by the random variable sequence $\tilde{X}_1^N$. How do we find a sequence at the output of the channel that is jointly typical with x^N.

Solution 4.13 In this example, the channel is not a symmetric channel and the input random variable is not uniformly distributed. The input probability mass function is given as

$$p(x=0) = \frac{3}{4}\, p(x=1) = \frac{1}{4}.$$

We can calculate the joint probability mass function using

$$p(x, y) = p(y|x)p(x)$$

as

$$\begin{aligned}
p(x=0, y=0) &= p(y=0|x=0)p(x=0) \rightarrow p(x=0, y=0) = 9/16 \\
p(x=0, y=1) &= p(y=1|x=0)p(x=0) \rightarrow p(x=0, y=1) = 3/16 \\
p(x=1, y=0) &= p(y=0|x=1)p(x=1) \rightarrow p(x=1, y=0) = 1/12 \\
p(x=1, y=1) &= p(y=1|x=1)p(x=1) \rightarrow p(x=1, y=1) = 2/12.
\end{aligned}$$

Either from Fig. 4.16, or from the joint probability mass function calculated, we can conclude that to find a sequence jointly typical with x^N, it is sufficient to flip $1/4$ of the total number of '0's in x^N and flip $1/3$ of the total number of '1's in x^N.

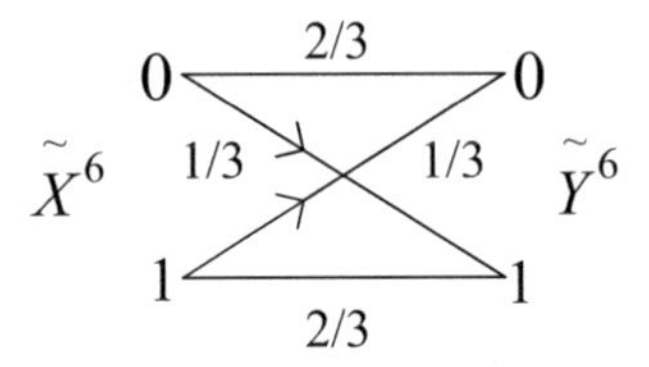

Fig. 4.17 Extended binary channel for Example 4.14

For instance, let $N = 36$, and $x^{36} = 000010011100001000011000001110000110$, we can find the sequence y^{36} jointly typical with x^{36} as

$$y^{36} = 01^*1^*01001111^*1^*1^*1^*100000^*0^*0000011100000^*0^*0$$

where $*$ indicates the flipped bits in the sequence x^{36}.

Example 4.14 For the discrete channel given in Fig. 4.17 we have $\tilde{X}^6 \sim p(x)$ and $p(x=0) = 1/2$, $p(x=1) = 1/2$. The generic random variable is $\tilde{X} \sim p(x)$.

(a) Find joint probability mass function $p(x, y)$.
(b) Find a jointly typical pair $\left(x^6, y^6\right)$ and mathematically verify its joint typicality.

Solution 4.14 (a) The joint probability mass function $p(x, y)$ can be calculated using

$$p(x, y) = p(y|x)p(x)$$

as

$$\begin{aligned} p(x=0, y=0) = \tfrac{1}{3}\; p(x=0, y=1) = \tfrac{1}{6} \\ p(x=1, y=1) = \tfrac{1}{3}\; p(x=1, y=0) = \tfrac{1}{6} \end{aligned}$$

(b) In strongly typical sequence x^N, '0' appears $N \times p(x=0)$ times and '1' appears $N \times p(x=0)$ times. Then, for $N = 6$, in the sequence x^6

$$\text{'0' appears } 6 \times \frac{1}{2} = 3 \text{ times}$$

$$\text{'1' appears } 6 \times \frac{1}{2} = 3 \text{ times.}$$

Thus, we can choose x^6 as "000111".

In strongly jointly typical sequences (x^N, y^N), the pair (x_i, y_i) occurs $N \times p(x_i, y_i)$ times. Then, for $N = 6$, in the jointly typical $\left(x^6, y^6\right)$

$$\begin{aligned}
&(x = 0, y = 0) \text{ appears} \rightarrow 6 \times p(0,0) \rightarrow 6 \times \frac{1}{3} \rightarrow 2 \text{ times} \\
&(x = 0, y = 1) \text{ appears} \rightarrow 6 \times p(0,1) \rightarrow 6 \times \frac{1}{6} \rightarrow 1 \text{ times} \\
&(x = 1, y = 0) \text{ appears} \rightarrow 6 \times p(1,0) \rightarrow 6 \times \frac{1}{6} \rightarrow 1 \text{ times} \\
&(x = 1, y = 1) \text{ appears} \rightarrow 6 \times p(1,1) \rightarrow 6 \times \frac{1}{3} \rightarrow 2 \text{ times}
\end{aligned} \tag{4.29}$$

Considering the information given in (4.29), we can decide on the jointly typical pair as in Fig. 4.18.

From Fig. 4.18, we see that $x^6 = 000111$ and $y^6 = 001011$ are jointly typical sequences.

Note that the found jointly typical pair is a strongly jointly typical pair.

Let's now mathematically verify that $x^6 = 000111$ and $y^6 = 001011$ are jointly typical sequences. For this purpose, we should show that

$$\left| -\frac{1}{6} \log p\left(x^6\right) - H\left(\tilde{X}\right) \right| < \epsilon \tag{4.30}$$

$$\left| -\frac{1}{6} \log p\left(y^6\right) - H\left(\tilde{Y}\right) \right| < \epsilon \tag{4.31}$$

and

$$\left| -\frac{1}{6} \log p\left(x^6, y^6\right) - H\left(\tilde{X}, \tilde{Y}\right) \right| < \epsilon. \tag{4.32}$$

The entropy $H\left(\tilde{X}\right)$ and $p\left(x^6\right)$ where $x^6 = 000111$ can be calculated as

$$H\left(\tilde{X}\right) = -\sum_x p(x) \log p(x) \rightarrow H\left(\tilde{X}\right) = -\left(\frac{1}{2} \log \frac{1}{2} + \frac{1}{2} \log \frac{1}{2}\right) \rightarrow H\left(\tilde{X}\right) = 1$$

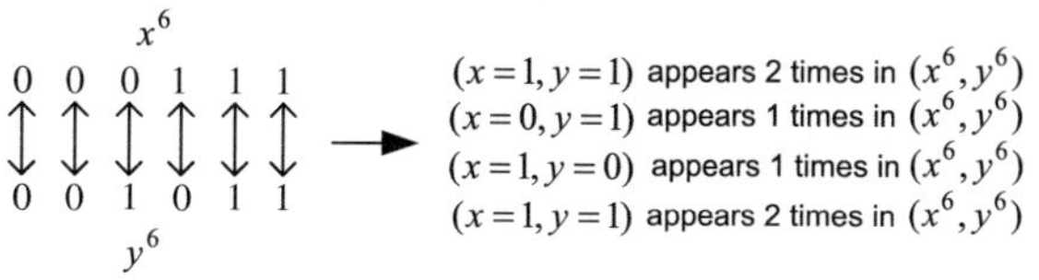

Fig. 4.18 Determination of jointly typical pairs

$$p(x^6) = p(x_1)p(x_2)\ldots p(x_6) \rightarrow p(x^6) = (p(x=0))^3 \times (p(x=1))^3 \rightarrow p(x^6) = \left(\frac{1}{2}\right)^6$$

Let $\epsilon = 0.01$, then (4.30) leads to

$$\left|-\frac{1}{6}\log p(x^6) - H(\tilde{X})\right| < \epsilon \rightarrow \left|-\frac{1}{6}\log\left(\frac{1}{2}\right) - 1\right| < 0.01 \rightarrow 0 < 0.01\surd$$

The entropy $H(\tilde{Y})$ and $p(y^6)$ where $y^6 = 001011$ can be calculated as

$$H(\tilde{Y}) = -\sum_y p(y)\log p(y) \rightarrow H(\tilde{Y}) = -\left(\frac{1}{2}\log\frac{1}{2} + \frac{1}{2}\log\frac{1}{2}\right) \rightarrow H(\tilde{Y}) = 1$$

$$p(y^6) = p(y_1)p(y_2)\ldots p(y_6) \rightarrow p(y^6) = (p(y=0))^3 \times (p(y=1))^3 \rightarrow p(y^6) = \left(\frac{1}{2}\right)^6$$

With $\epsilon = 0.01$, then (4.31) leads to

$$\left|-\frac{1}{6}\log p(y^6) - H(\tilde{Y})\right| < \epsilon \rightarrow \left|-\frac{1}{6}\log\left(\frac{1}{2}\right) - 1\right| < 0.01 \rightarrow 0 < 0.01\surd$$

The joint entropy $H(\tilde{X},\tilde{Y})$ and $p(x^6, y^6)$ where $x^6 = 000111, y^6 = 001011$ can be calculated as

$$H(\tilde{X},\tilde{Y}) = -\sum_{x,y} p(x,y)\log p(x,y) \rightarrow H(\tilde{X},\tilde{Y}) = -\left(\frac{2}{3}\log\frac{1}{3} + \frac{2}{6}\log\frac{1}{6}\right) \rightarrow H(\tilde{X},\tilde{Y}) = 1.92$$

$$p(x^6, y^6) = p(x_1, y_1)p(x_2, y_2)\ldots p(x_6, y_6) \rightarrow p(x^6, y^6) = p(0,0)^2 p(1,1)^2 p(0,1)p(1,0) \rightarrow p(x^6, y^6) = \left(\frac{1}{3}\right)^4\left(\frac{1}{6}\right)^2$$

With $\epsilon = 0.01$, then (4.32) leads to

$$\left|-\frac{1}{6}\log p(x^6, y^6) - H(\tilde{X},\tilde{Y})\right| < \epsilon \rightarrow \left|-\frac{1}{6}\log\left(\frac{1}{3}\right)^4\left(\frac{1}{6}\right)^2 - 1.92\right| < 0.01 \rightarrow 0.0017 < 0.01\surd$$

Since, all the three (4.30), (4.31), and (4.32) are satisfied, then $\left(x^6, y^6\right)$ is jointly typical.

4.3.3 *Number of Jointly Typical Sequences and Probability for Typical Sequences*

Consider the random variable sequences $(\tilde{X}^N, \tilde{Y}^N)$ which produces the real number sequences (x^N, y^N) whose joint probability mass function satisfy

$$p\left(x^N, y^N\right) = \prod_{i=1}^{N} p(x_i, y_i). \tag{4.33}$$

x^N and y^N are typical sequences. Now, we ask the question: How many jointly typical sequences (x^N, y^N) do we have?

Let $p_i = p(x_i, y_i)$, i.e., $p_1 = p(x_1, y_1), p_2 = p(x_2, y_2), \ldots, p_N = p(x_N, y_N)$. As $N \to \infty$, the pair (x_1, y_1) appears $N \times p_1$ times in the sequence (x^N, y^N), and the pair (x_2, y_2) appears $N \times p_2$ times in the sequence (x^N, y^N) and so on. Then, the probability of the sequence (x_N, y_N) can be calculated as

$$p\left(x^N, y^N\right) = p_1^{N \times p_1} \times p_2^{N \times p_2} \times \cdots \times p_N^{N \times p_N} \tag{4.34}$$

which can be written as

$$p\left(x^N, y^N\right) = 2^{\log p_1^{N \times p_1}} \times 2^{\log p_2^{N \times p_2}} \times \cdots \times 2^{\log p_N^{N \times p_N}} \tag{4.35}$$

which can be expressed as

$$p\left(x^N, y^N\right) = 2^{N \times \sum_{i=1}^{N} p_i \log p_i} \tag{4.36}$$

where

$$\sum_{i=1}^{N} p_i \log p_i$$

is nothing but $-H(\tilde{X}, \tilde{Y})$. Thus, (4.35) can be written as

$$p\left(x^N, y^N\right) = 2^{-N \times H(\tilde{X}, \tilde{Y})}. \tag{4.37}$$

If every jointly typical sequence (x^N, y^N) has the same probability, then from (Eq. 4.37) we can conclude that the number of jointly typical sequences equals to

$$\frac{1}{p(x^N, y^N)} \to 2^{N \times H(\tilde{X}, \tilde{Y})}. \tag{4.38}$$

For the derivation of (4.38) we started with the strongly jointly typical assumption in (4.34), however, the derived formula (4.38) represents the total number of jointly typical sequences including strongly and weakly jointly typical sequences. The reason for this result is that different pairs such as (0,1) and (1,0) can have the same probability, and these probabilities are used in (4.34) without accounting their different corresponding pairs.

Example 4.15 For the binary communication channel shown in Fig. 4.19 where $\tilde{X}_1^{20} \sim p(x)$, $\tilde{Y}_1^{20} \sim p(y)$ and we have the generic random variables $\tilde{X} \sim p(x), \tilde{Y} \sim p(y)$ with range sets $R_{\tilde{X}} = \{0, 1\}$, $R_{\tilde{Y}} = \{0, 1\}$, and $p(x = 0) = 3/4$, $p(x = 1) = 1/4$.
Find the number of jointly typical sequences (x^{20}, y^{20}).

Solution 4.15 In Example 4.13, we calculated the joint probability mass function as

$$\begin{aligned}
p(x = 0, y = 0) &= p(y = 0|x = 0)p(x = 0) \to p(x = 0, y = 0) = 9/16 \\
p(x = 0, y = 1) &= p(y = 1|x = 0)p(x = 0) \to p(x = 0, y = 1) = 3/16 \\
p(x = 1, y = 0) &= p(y = 0|x = 1)p(x = 1) \to p(x = 1, y = 0) = 1/12 \\
p(x = 1, y = 1) &= p(y = 1|x = 1)p(x = 1) \to p(x = 1, y = 1) = 2/12.
\end{aligned}$$

The joint entropy $H(\tilde{X}, \tilde{Y})$ can be calculated using

$$H(\tilde{X}, \tilde{Y}) = \sum_{x,y} p(x, y) \log p(x, y)$$

as

$$\begin{aligned}
H(\tilde{X}, \tilde{Y}) &= \frac{9}{16} \times \log \frac{9}{16} + \frac{3}{16} \times \log \frac{3}{16} + \frac{1}{12} \times \log \frac{1}{12} + \frac{2}{12} \times \log \frac{2}{12} \to H(\tilde{X}, \tilde{Y}) \\
&= 1.6493.
\end{aligned}$$

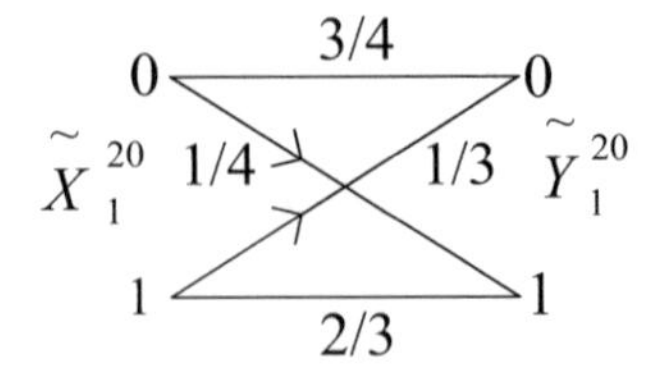

Fig. 4.19 Extended binary communication channel for Example 4.15

The number of jointly typical sequences equals to

$$2^{N\times H(\tilde{X},\tilde{Y})} \rightarrow 2^{20\times 1.6493} \rightarrow 8.5070 \times 10^{9}.$$

The Probability of Selecting a Jointly Typical Sequence from a Number of Typical Sequence Pairs (x^N, y^N):

We assume that x^N and y^N are typical sequences, unless otherwise indicated.

The number of y^N *sequences each of which is jointly typical with* x^N:
Considering all the pairs (x^N, y^N), we know that

$$2^{N\times H(\tilde{X},\tilde{Y})} \tag{4.39}$$

of these pairs are jointly typical sequences. Then, for a single typical sequence x^N there are a number of typical sequences y^N each of which is jointly typical with x^N. The number of such y^N sequences in average can be calculated as

$$\frac{2^{N\times H(\tilde{X},\tilde{Y})}}{2^{NH(\tilde{X})}} \rightarrow 2^{N\times\left(H(\tilde{X},\tilde{Y})-H(\tilde{X})\right)} \rightarrow 2^{N\times\left(H(\tilde{X},\tilde{Y})-H(\tilde{X})\right)} \rightarrow 2^{N\times\left(H(\tilde{Y}|\tilde{X})\right)}$$

Thus, for the typical sequence x^N, there are $2^{N\times\left(H(\tilde{Y}|\tilde{X})\right)}$ number of y^N sequences in average such that each of them is jointly typical with x^N. Let's call the set of $2^{N\times\left(H(\tilde{Y}|\tilde{X})\right)}$ number of y^N sequences as typical range set.

The total number of typical y^N sequences equals to $2^{NH(\tilde{Y})}$. The number of typical range sets can be calculated as

$$\frac{2^{NH(\tilde{Y})}}{2^{N\times\left(H(\tilde{Y}|\tilde{X})\right)}} \rightarrow 2^{N\times I(\tilde{X},\tilde{Y})}.$$

The concept of typical range set is depicted in Fig. 4.20.

Note: The number of sequences in typical range sets are equal to each other for binary symmetric channel. However, it may be different for binary non-symmetric channel.

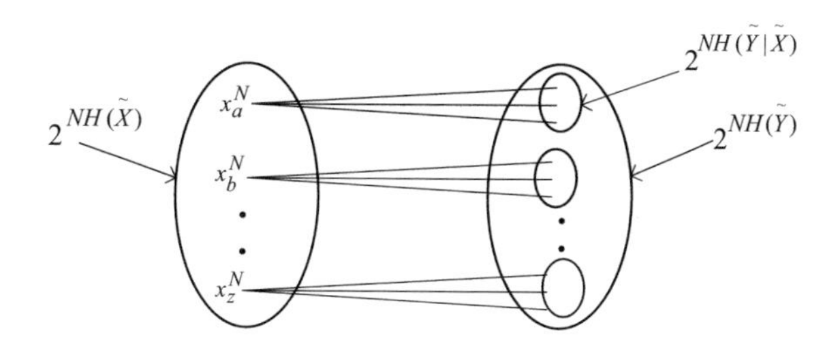

Fig. 4.20 The concept of typical range set

The probability of (x^N, y^N) *to be jointly typical*:
The number of typical sequences x^N equals to

$$2^{NH(\tilde{X})}$$

and the number of typical sequences y^N equals to

$$2^{NH(\tilde{Y})}.$$

If we consider the pair of sequences (x^N, y^N), it is obvious that there are

$$2^{NH(\tilde{X})} \times 2^{NH(\tilde{Y})} \to 2^{N\left(H(\tilde{X})+H(\tilde{Y})\right)} \tag{4.40}$$

number of pairs available. Equation (4.40) can be considered as the number of elements of the Cartesian product of two sets.
Now let's consider the following question. Assume that x^N and y^N are separately typical sequences, then what is the probability that they are jointly typical as well? We can calculate the probability of (x^N, y^N) to be jointly typical by taking the division of (4.39) and (4.40) as

$$\frac{2^{N\times H(\tilde{X},\tilde{Y})}}{2^{N\left(H(\tilde{X})+H(\tilde{Y})\right)}} \to 2^{N\times\left(H(\tilde{X},\tilde{Y})-H(\tilde{X})-H(\tilde{Y})\right)} \to 2^{-N\times I(\tilde{X},\tilde{Y})}. \tag{4.41}$$

Equation (4.41) implies that if x^N is a typical sequence, to come across a typical sequence y^N which is jointly typical with x^N we should check at least $2^{N\times I(\tilde{X},\tilde{Y})}$ number of y^N sequences.

On the other hand, if we consider all the sequences, i.e., typical or non-typical, the number of x^N sequences equals to $|R_{\tilde{X}}|^N$ and the number of y^N sequences equals to $|R_{\tilde{Y}}|^N$. Then, the probability of (x^N, y^N) to be jointly typical equals to

$$\frac{2^{N\times H(\tilde{X},\tilde{Y})}}{|R_{\tilde{X}}|^N \times |R_{\tilde{Y}}|^N}. \tag{4.42}$$

Example 4.16 Consider the extended binary symmetric channel shown in Fig. 4.21 where $\tilde{X}_1^{20} \sim p(x)$, $\tilde{Y}_1^{20} \sim p(y)$ and we have the generic random variables $\tilde{X} \sim p(x), \tilde{Y} \sim p(y)$ with range sets $R_{\tilde{X}} = \{0, 1\}$, $R_{\tilde{Y}} = \{0, 1\}$, and $p(x = 0) = 1/2$, $p(x = 1) = 1/2$.

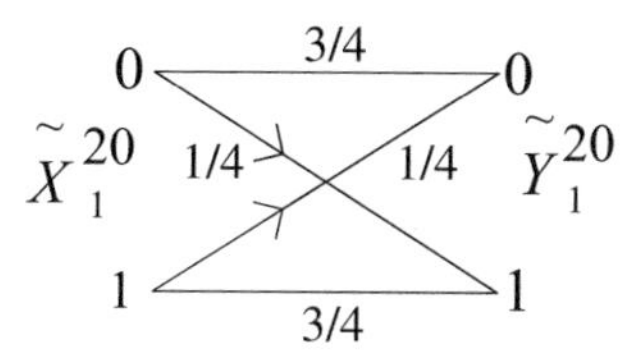

Fig. 4.21 Extended binary communication channel for Example 4.16

(a) How many x^{20} typical sequences are there?
(b) How many y^{20} typical sequences are there?
(c) Consider a single typical sequence x^{20}, how many y^{20} typical sequences each of which is jointly typical with x^{20} are there?
(d) x^{20} is a typical sequence, y^{20} is a typical sequence. What is the probability that (x^{20}, y^{20}) is jointly typical?

Solution 4.16 The entropies $H(\tilde{X})$ and $H(\tilde{Y})$ can be calculated as

$$H(\tilde{X}) = 1, \quad H(\tilde{Y}) = 1.$$

(a) The number of x^{20} typical sequences is

$$2^{20 \times H(\tilde{X})} \rightarrow 2^{20}.$$

(b) The number of y^{20} typical sequences is

$$2^{20 \times H(\tilde{Y})} \rightarrow 2^{20}.$$

(c) The conditional entropy $H(\tilde{Y}|\tilde{X})$ can be calculated using

$$H(\tilde{Y}|\tilde{X}) = -\sum_{x,y} p(x,y) \log p(y|x) \rightarrow H(\tilde{Y}|\tilde{X}) = -\sum_{x,y} p(y|x)p(x) \log p(y|x)$$

as

$$H(\tilde{Y}|\tilde{X}) = -\left(2 \times \frac{3}{8} \log \frac{3}{4} + 2 \times \frac{1}{8} \log \frac{1}{4}\right) \rightarrow H(\tilde{Y}|\tilde{X}) = 0.8113.$$

The number of typical sequences y^{20} which are jointly typical with x^{20} equals to

$$2^{20 \times H(\tilde{Y}|\tilde{X})} \rightarrow 2^{20 \times 0.8113} \rightarrow 76650.$$

(d) The probability that (x^{20}, y^{20}) is jointly typical equals to

$$\frac{2^{20\times H(\tilde{X},\tilde{Y})}}{2^{20(H(\tilde{X})+H(\tilde{Y}))}} \to 2^{20\times(H(\tilde{Y}|\tilde{X})-H(\tilde{Y}))} \to 2^{20\times(0.8113-1)} \to 0.0731.$$

Exercise Repeat the previous example for the channel given in Fig. 4.22 where $\tilde{X}_1^{20} \sim p(x)$, $\tilde{Y}_1^{20} \sim p(y)$ and we have the generic random variables $\tilde{X} \sim p(x), \tilde{Y} \sim p(y)$ with range sets $R_{\tilde{X}} = \{0,1\}$, $R_{\tilde{Y}} = \{0,1\}$, and $p(x=0) = 3/4$, $p(x=1) = 1/4$.
Let (x^N, y^N) be the sequences generated by the random variable sequences $(\tilde{X}^N, \tilde{Y}^N)$ according to the joint probability mass function $p(x^N, y^N)$ which satisfy

$$p(x^N, y^N) = \prod_{i=1}^{N} p(x_i, y_i).$$

Then, we have the following theorems.

Theorem 4.1 *Prob*$(x^N, y^N)\epsilon T_\epsilon^N \to 1$ *as N goes to infinity.*

Proof 4.1 The proof comes directly from the definition of jointly typical sequences. According to the weak law of large numbers and from the definition, we have

$$Prob\left(\left|-\frac{1}{N}\log p(x) - H(\tilde{X})\right| > \epsilon\right) < \epsilon_1, \quad \text{where } N > N_a$$

$$Prob\left(\left|-\frac{1}{N}\log p(y) - H(\tilde{Y})\right| > \epsilon\right) < \epsilon_1, \quad \text{where } N > N_b$$

$$Prob\left(\left|-\frac{1}{N}\log p(x,y) - H(\tilde{X},\tilde{Y})\right| > \epsilon\right) < \epsilon_1, \quad \text{where } N > N_c.$$

Choosing $N \geq \max(N_a, N_b, N_c)$, we get

$$Prob\left((x^N, y^N)\epsilon T_\epsilon^N\right) = Prob\begin{pmatrix} \left|-\frac{1}{N}\log p(x^N) - H(\tilde{X})\right| > \epsilon \\ \cup\left|-\frac{1}{N}\log p(y^N) - H(\tilde{Y})\right| > \epsilon \\ \cup\left|-\frac{1}{N}\log p(x^N, y^N) - H(\tilde{X},\tilde{Y})\right| > \epsilon \end{pmatrix} < \epsilon$$

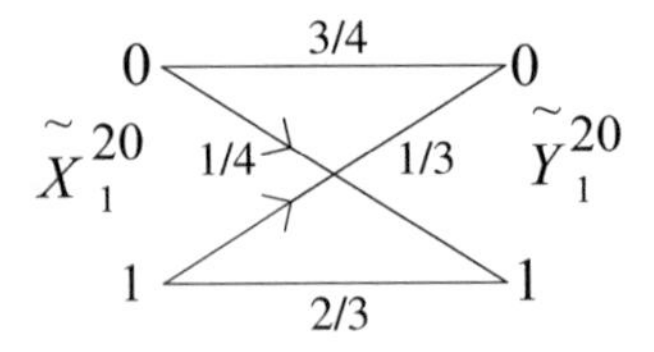

Fig. 4.22 Extended binary communication channel for exercise

which implies that

$$Prob\left(x^N, y^N\right) \epsilon T_\epsilon^N \rightarrow 1.$$

Theorem 4.2 $\left|T_\epsilon^N\right| \leq 2^{N\times\left(H\left(\tilde{X},\tilde{Y}\right)+\epsilon\right)}$ *where* $\left|T_\epsilon^N\right|$ *shows the number of jointly typical sequences* (x^N, y^N) *in the typical set* T_ϵ^N.

Proof 4.2 The sum of the probabilities of the jointly typical sequences gives us

$$\sum_{T_{x,y}^N} p(x^N, y^N) = 1$$

which can be written as

$$\sum_{T_\epsilon^N} p(x^N, y^N) \leq 1$$

in which employing

$$2^{-N\left(\left(H\left(\tilde{X},\tilde{Y}\right)+\epsilon\right)\right)} \leq p(X^N, Y^N)$$

we get

$$\sum_{T_\epsilon^N} 2^{-N\left(\left(H\left(\tilde{X},\tilde{Y}\right)+\epsilon\right)\right)} \leq 1$$

leading to

$$\left|T_\epsilon^N\right| \leq 2^{N\times\left(H\left(\tilde{X},\tilde{Y}\right)+\epsilon\right)}.$$

Theorem 4.3 $\left|T_\epsilon^N\right| \geq (1-\epsilon)2^{N\times\left(H\left(\tilde{X},\tilde{Y}\right)-\epsilon\right)}$ *where* $\left|T_\epsilon^N\right|$ *shows the number of jointly typical sequences* (x^N, y^N) *in the typical set* T_ϵ^N.

Proof 4.3 For a very large N value we have

$$Prob\left(T_\epsilon^N\right) \geq 1-\epsilon$$

which implies that

$$\sum_{T_\epsilon^N} p(x^N, y^N) \geq 1-\epsilon$$

where employing

$$2^{-N\times\left(H\left(\tilde{X},\tilde{Y}\right)-\epsilon\right)}\geq p(x^N,y^N)$$

we obtain

$$\sum_{T_\epsilon^N} 2^{-N\times\left(H\left(\tilde{X},\tilde{Y}\right)-\epsilon\right)}\geq 1-\epsilon$$

leading to

$$\left|T_\epsilon^N\right|\geq(1-\epsilon)2^{N\times\left(H\left(\tilde{X},\tilde{Y}\right)-\epsilon\right)}. \tag{4.43}$$

Theorem 4.4 *Let* $(\tilde{A}^N,\tilde{B}^N)$ *be the random variable sequence having the joint probability mass function* $p(a^N,b^N)$ *satisfying*

$$p\left(a^N,b^N\right)=p(a^N)p(b^N)$$

i.e., $\tilde{A}^N$ *and* $\tilde{B}^N$ *are independent sequences. The random variables sequences* $\tilde{A}^N$ *and* $\tilde{B}^N$ *have the same marginal as* $\tilde{X}^N$ *and* $\tilde{Y}^N$, i.e.,

$$Prob\left(\tilde{A}^N=x^N\right)=Prob\left(\tilde{X}^N=x^N\right)$$
$$Prob\left(\tilde{B}^N=y^N\right)=Prob\left(\tilde{B}^N=y^N\right)$$

Then, we have

$$Prob\left(\left(a^N,b^N\right)\epsilon\, T_\epsilon^N\right)\leq 2^{-N\times\left(I\left(\tilde{X},\tilde{Y}\right)-\epsilon\right)} \tag{4.44}$$

And

$$Prob\left(\left(a^N,b^N\right)\epsilon\, T_\epsilon^N\right)\geq(1-\epsilon)2^{-N\times\left(I\left(\tilde{X},\tilde{Y}\right)+\epsilon\right)}. \tag{4.45}$$

It is clear from (4.44) *to* (4.45) *that as* $N\to\infty$, *we get*

$$Prob\left(\left(a^N,b^N\right)\epsilon\, T_\epsilon^N\right)\to 0.$$

This theorem implies that for the randomly chosen typical sequences x^N *and* y^N *the probability that the chosen pair* (x^N,y^N) *is jointly typical approximately equals to*

$$2^{-N\times I\left(\tilde{X},\tilde{Y}\right)}.$$

In fact, we already gave this result in (4.41).

Proof 4.4

$$Prob\left(\left(\tilde{A}^N = x^N, \tilde{B}^N = y^N\right)\epsilon\, T_\epsilon^N\right) = \sum_{(x^N, y^N)\epsilon\, T_\epsilon^N} p\left(x^N\right)p\left(y^N\right)$$

in which employing

$$p\left(x^N\right) \le 2^{-N\times\left(H\left(\tilde{X}\right)-\epsilon\right)} p\left(y^N\right) \le 2^{-N\times\left(H\left(\tilde{Y}\right)-\epsilon\right)}$$

we get

$$Prob\left(\left(\tilde{A}^N, \tilde{B}^N\right)\epsilon\, T_\epsilon^N\right) \le \sum_{(x^N, y^N)\epsilon T_\epsilon^N} 2^{-N\times\left(H\left(\tilde{X}\right)-\epsilon\right)} 2^{-N\times\left(H\left(\tilde{Y}\right)-\epsilon\right)}$$

which can be written as

$$Prob\left(\left(\tilde{A}^N, \tilde{B}^N\right)\epsilon\, T_\epsilon^N\right) \le 2^{-N\times\left(H\left(\tilde{X}\right)-\epsilon\right)} 2^{-N\times\left(H\left(\tilde{Y}\right)-\epsilon\right)} \sum_{(x^N, y^N)\epsilon T_\epsilon^N} 1$$

which is evaluated as

$$Prob\left(\left(\tilde{A}^N, \tilde{B}^N\right)\epsilon\, T_\epsilon^N\right) \le 2^{N\times\left(H\left(\tilde{X},\tilde{Y}\right)+\epsilon\right)} 2^{-N\times\left(H\left(\tilde{X}\right)-\epsilon\right)} 2^{-N\times\left(H\left(\tilde{Y}\right)-\epsilon\right)}$$

leading to

$$Prob\left(\left(\tilde{A}^N, \tilde{B}^N\right)\epsilon T_\epsilon^N\right) \le 2^{-N\times\left(I\left(\tilde{X},\tilde{Y}\right)-\epsilon\right)}.$$

To prove (4.45), we have

$$Prob\left(\left(\tilde{A}^N, \tilde{B}^N\right)\epsilon T_\epsilon^N\right) = \sum_{(x^N, y^N)\epsilon T_\epsilon^N} p\left(x^N\right)p\left(y^N\right)$$

in which employing

$$p\left(x^N\right) \ge 2^{-N\times\left(H\left(\tilde{X}\right)+\epsilon\right)} p\left(y^N\right) \ge 2^{-N\times\left(H\left(\tilde{Y}\right)+\epsilon\right)}$$

and

$$\left|T_\epsilon^N\right| \ge (1-\epsilon)2^{N\times\left(H\left(\tilde{X},\tilde{Y}\right)-\epsilon\right)}$$

we obtain

$$Prob\left(\left(a^N, b^N\right)\epsilon T_\epsilon^N\right) \geq (1-\epsilon)2^{-N\times\left(I\left(\tilde{X},\tilde{Y}\right)+\epsilon\right)}.$$

Note: ϵ represents a very small number, so, $k \times \epsilon$ is also a very small number, then we can use ϵ for the place of $k \times \epsilon$.

4.4 Channel Coding Theorem

Before stating the channel coding theorem, let's solve some examples to prepare ourselves for the proof of the channel coding theorem.

Example 4.17 A communication channel is depicted in the Fig. 4.23.
If N is a very large number, then the sequence x^N at the input of the channel becomes a typical sequence, and when the typical sequence is passed through the channel we get typical sequence y^N at the output of the channel.
What is the probability that x^N and y^N are not jointly typical sequences?

Solution 4.17 From Theorem 4.1, we can say that for very large N values, the probability that x^N and y^N are jointly typical sequences approaches '1', i.e.,

$$Prob\left(x^N \text{ and } y^N \text{are jointly typical}\right) \to 1$$

Then, for very large N values

$$Prob\left(x^N \text{ and } y^N \text{are not jointly typical}\right) \to 0$$

Example 4.18 For the previous example, let w^N be another typical sequence at the input of the channel. What is the probability that w^N is jointly typical with y^N?

Solution 4.18 According to Theorem 4.4, the probability of any two typical sequences w^N and y^N to be jointly typical is bounded by

$$Prob\left(w^N \text{ is jointly typical with } y^N\right) \leq 2^{-N\times\left(I\left(\tilde{W};\tilde{Y}\right)-\epsilon\right)}.$$

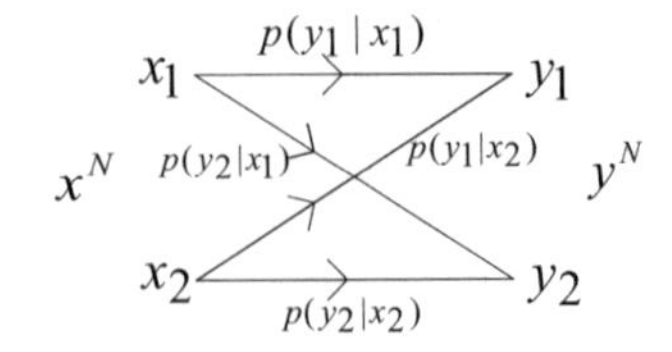

Fig. 4.23 A communication channel for Example 4.17

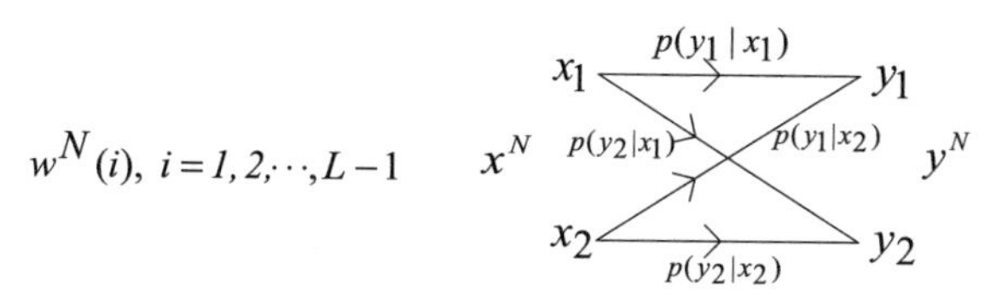

Fig. 4.24 A communication channel for Example 4.19

Example 4.19 A communication channel is depicted in the Fig. 4.24.
If N is a very large number, then the sequence x^N at the input of the channel becomes a typical sequence, and when the typical sequence is passed through the channel we get typical sequence y^N at the output of the channel. According to Theorem 4.1 the sequences x^N and y^N are jointly typical. Assume that we have a group of typical sequences indicated by

$$w^N(i), i = 1, \ldots, L-1$$

where $w^N(i)$ represents

$$w^N(i) = w_{i1}w_{i2}\ldots w_{iN}.$$

For instance,

$$w^N(1) = w_{11}w_{12}\ldots w_{1N}.$$

What is the probability that y^N is jointly typical with one of the sequences available in the group $w^N(i), i = 1, \ldots, L-1$.

Solution 4.19 According to Theorem 4.4, the probability of any two typical sequences $w(i)^N$ and y^N to be jointly typical is bounded by

$$Prob\left(w(i)^N \text{ is jointly typical with } y^N\right) \leq 2^{-N\times\left(I(\tilde{W};\tilde{Y})-\epsilon\right)} \tag{4.46}$$

And if we consider all the sequences in the group, we can write that

$$\begin{aligned}
&Prob\left(y^N \text{ is jointly typical with one of } w(i)^N, i = 1, \ldots, L-1\right) = \\
&+Prob\left(y^N \text{ is jointly typical with one of } w(1)^N\right) \\
&+Prob\left(y^N \text{ is jointly typical with one of } w(2)^N\right) \\
&+Prob\left(y^N \text{ is jointly typical with one of } w(3)^N\right) \\
&+Prob\left(y^N \text{ is jointly typical with one of } w(L-1)^N\right)
\end{aligned} \tag{4.47}$$

Using (4.46) for (4.47), we get

$$Prob\left(y^N \text{is jointly typical with one of } w(i)^N, i = 1, \ldots, L-1\right) \leq (L-1)2^{-N\times\left(I\left(\tilde{W};\tilde{Y}\right)-\epsilon\right)}$$

which can also be written as

$$Prob\left(w(i)^N, i = 1, \ldots, L-1 \text{ is jointly typical with } y^N\right) \leq L2^{-N\times\left(I\left(\tilde{W};\tilde{Y}\right)-\epsilon\right)} \tag{4.48}$$

If the sequences $w(i)^N$ are generated by an encoder of rate $R = K/N$, then we have

$$L = 2^{NR}. \tag{4.49}$$

Substituting (4.49) into (4.48) we get,

$$Prob\left(w(i)^N, i = 1, \ldots, L-1 \text{ is jointly typical with } y^N\right) \leq 2^{NR}2^{-N\times\left(I\left(\tilde{W};\tilde{Y}\right)-\epsilon\right)}$$

which can be simplified as

$$Prob\left(w(i)^N, i = 1, \ldots, L-1 \text{ is jointly typical with } y^N\right) \leq 2^{-N\times\left(I\left(\tilde{W};\tilde{Y}\right)-R-\epsilon\right)}.$$

Now it is time to state the channel coding theorem again, and make its proof.

Channel Coding Theorem

For a discrete memoryless channel with capacity C as shown in Fig. 4.25, there exists a channel code (N, K) with rate $R < C$ such that the maximal probability error of the code, i.e., β^*, goes to zero as N goes to infinity, i.e.,

$$\text{If } R < C, \text{ then } \beta^* \to 0 \text{ as } N \to \infty$$

Proof In Examples 4.17, 4.18 and 4.19, we provided parts of the proof as exercises. In Fig. 4.25, $x^N(m)$ is the code-word at the channel input for the message index m and y^N is the channel output. The decoder searches the code words $x^N(z)$ and

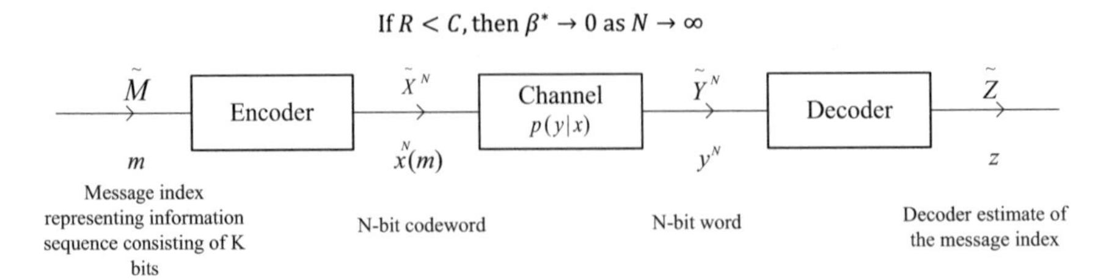

Fig. 4.25 A typical communication system

finds the one that is jointly typical with y^N and declares the index of the code-word that is jointly typical with y^N as the transmitted message index m.
The decoder makes two types of errors while making its decision. These errors are:

(a) The decoder decides that the code-word $x^N(m)$ is not jointly typical with y^N. Let's denote this event by E^c.
(b) The decoder decides that another code-word $x^N(\hat{m})$ is jointly typical with y^N, thus outputs wrong index value $z = \hat{m}$. Let's denote this event by $E_{\hat{m}}$.

The probability of decoding error can be written as

$$P_e \leq Prob(E^c) + \sum_{\substack{\hat{m} \\ \hat{m} \neq m}} Prob(E_{\hat{m}}) \tag{4.50}$$

in which using Theorem 4.1, we can write that

$$Prob(E^c) \rightarrow 0$$

and using Theorem 4.4, $Prob(E_{\hat{m}})$ can be bounded as

$$Prob(E_{\hat{m}}) \leq 2^{-N \times \left(I\left(\tilde{X};\tilde{Y}\right) - \epsilon\right)}.$$

If we consider all the code-words other than the transmitted one, (4.50) can be written as

$$P_e \leq (M-1) 2^{-N \times \left(I\left(\tilde{X};\tilde{Y}\right) - \epsilon\right)} \tag{4.51}$$

in which employing $M = 2^{NR}$, we obtain

$$P_e \leq 2^{-N \times \left(I\left(\tilde{W};\tilde{Y}\right) - R - \epsilon\right)} \tag{4.52}$$

For the expression in (4.52), if

$$I\left(\tilde{X};\tilde{Y}\right) - R > 0 \rightarrow R < I\left(\tilde{X};\tilde{Y}\right) \tag{4.53}$$

as $N \rightarrow \infty$ we get

$$P_e \rightarrow 0.$$

If maximum value of $I\left(\tilde{X};\tilde{Y}\right)$ is considered, i.e., capacity is considered, then (4.53) can be expressed as, if

$$R < C \tag{4.54}$$

as $N \to \infty$ we get

$$P_e \to 0. \tag{4.55}$$

This means for very large code-word lengths, if the rate of the code is smaller than the capacity of the channel, then probability of the decoding error goes to zero, i.e., it is possible to find the transmitted message index.

Converse of the Channel Coding Theorem:

If the maximal probability of error of a channel code (*N*, *K*) goes to zero as *N* goes to infinity, then we must have $R < C$, i.e., for an (*N*, *K*) channel code

$$\text{If } \beta^* \to 0, \text{ as } N \to \infty, \text{ then we must have } R < C.$$

Proof Considering the communication system in Fig. 4.25, it is seen that we have the Markov chain

$$\tilde{M} \to \tilde{X}^N \to \tilde{Y}^N \to \tilde{Z}.$$

Using Fano's inequality we can write

$$H\left(\tilde{M}|\tilde{Z}\right) \leq 1 + p_N \log 2^{NR} = 1 + p_N NR \tag{4.56}$$

where p_N is the probability of decoding error. Since the message is uniformly distributed, we have

$$H\left(\tilde{M}\right) = NR. \tag{4.57}$$

The entropy

$$H\left(\tilde{M}\right)$$

can be written as

$$H\left(\tilde{M}\right) = H\left(\tilde{M}|\tilde{Z}\right) + I\left(\tilde{M},\tilde{Z}\right)$$

where using the Fano's inequality in (4.56), and employing the data processing inequality for $I(\tilde{M}, \tilde{Z})$, i.e.,

$$I(\tilde{M}, \tilde{Z}) \leq I(\tilde{M}, \tilde{Y}^N)$$

we get

$$H(\tilde{M}) \leq 1 + p_N NR + I(\tilde{M}, \tilde{Y}^N)$$

in which expanding $I(\tilde{M}, \tilde{Y}^N)$ as

$$I(\tilde{M}, \tilde{Y}^N) = \sum_{i=1}^{N} I(\tilde{M}; \tilde{Y}_i | \tilde{Y}^{i-1})$$

we obtain

$$H(\tilde{M}) \leq 1 + p_N NR + \sum_{i=1}^{N} I(\tilde{M}; \tilde{Y}_i | \tilde{Y}^{i-1})$$

which can be written as

$$H(\tilde{M}) \leq 1 + p_N NR + \sum_{i=1}^{N} I(\tilde{M}, \tilde{Y}^{i-1}; \tilde{Y}_i)$$

where again employing the data processing inequality, we obtain

$$H(\tilde{M}) \leq 1 + p_N NR + \sum_{i=1}^{N} I(\tilde{X}_i; \tilde{Y}_i)$$

leading to

$$H(\tilde{M}) \leq 1 + p_N NR + \sum_{i=1}^{N} \max_{p(x)} I(\tilde{X}, \tilde{Y})$$

which can be simplified as

$$H(\tilde{M}) \leq 1 + p_N NR + N \max_{p(x)} I(\tilde{X}, \tilde{Y}). \tag{4.58}$$

Considering (4.57) and (4.58) together, we can write

$$NR \leq 1 + p_N NR + N \max_{p(x)} I(\tilde{X}, \tilde{Y})$$

from which, we get

$$R \leq \frac{1}{N} + p_N R + \max_{p(x)} I(\tilde{X}, \tilde{Y})$$

which implies that as $p_N \to 0$, we obtain

$$R \leq \max_{p(x)} I(\tilde{X}, \tilde{Y}).$$

Problems

1. For the communication system given in Fig. 4.P1 which code rates can be achievable.
2. For the communication system given in Fig. 4.P2 $K = 2$, $N = 3$. Design a simple encoding method and choose a decoding function $d(\cdot)$ for the decoder, and calculate the conditional probability of error for the code-words.
3. Using the discrete channel given in Fig. 4.P3 where $\tilde{X}_1^{36} \sim p(x)$, $\tilde{Y}_1^{360} \sim p(y)$ and we have the generic random variables $\tilde{X} \sim p(x)$, $\tilde{Y} \sim p(y)$ with range sets $R_{\tilde{X}} = \{0, 1\}$, $R_{\tilde{Y}} = \{0, 1\}$, and $p(x = 0) = 1/4$, $p(x = 1) = 3/4$.

(a) Find the probability mass function $p(y)$.
(b) How many x^{80} typical sequences are there?
(c) How many y^{80} typical sequences are there?
(d) Consider a single typical sequence x^{80}, how many y^{80} typical sequences each of which is jointly typical with x^{20} are there?

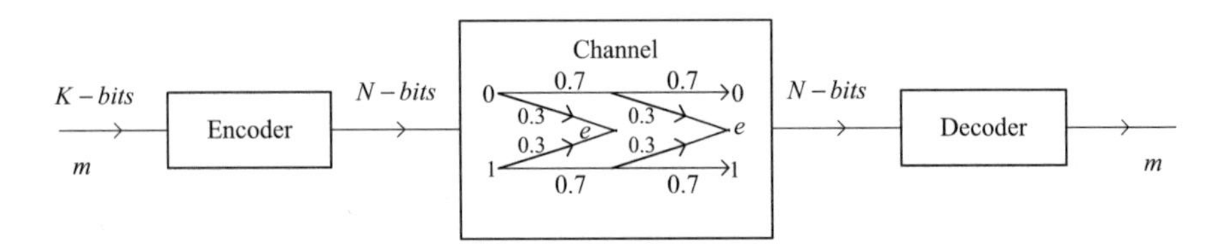

Fig. 4.P1 Communication system employing concatenated binary erasure channel

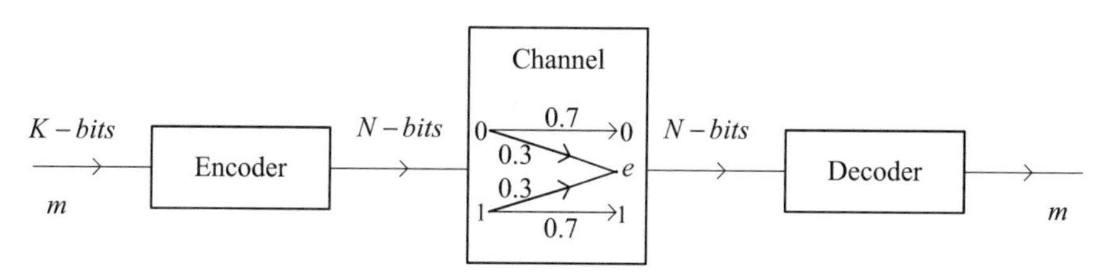

Fig. 4.P2 Communication system employing binary erasure channel

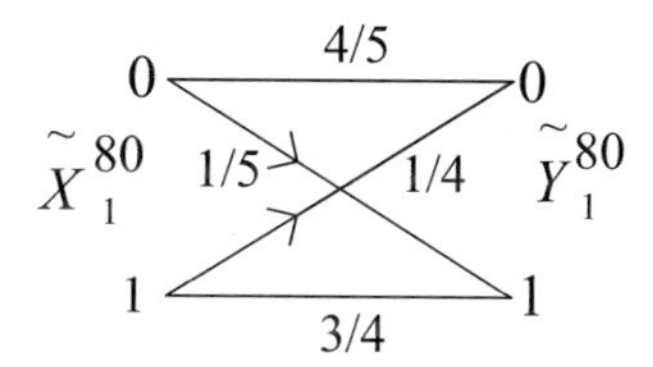

Fig. 4.P3 Binary non-symmetric channel

(e) x^{80} is a typical sequence, y^{80} is a typical sequence. What is the probability that (x^{80}, y^{80}) is jointly typical?

(f) How can you find a strongly typical sequence x^{20}?

(g) How can you find a strongly typical sequence y^{20}?

(h) Find joint probability mass function $p(x, y)$.

References

1. Cover TM, Thomas JA (2006) Elements of information theory 2nd Edn. ISBN-10: 0471241954
2. MacKay DJC (2003) Information theory, inference and learning algorithms hardcover, Cambridge University. ISBN-10: 0521642981
3. Gray RM (2011) Entropy and information theory, hardcover, Springer, US. ISBN: 978-1-4419-7969-8
4. Fazlollah MR (1995) An introduction to information theory, Dover books on mathematics. ISBN-10: 0486682102

O. Gazi, *Information Theory for Electrical Engineers*, Signals and Communication Technology, https://doi.org/10.1007/978-981-10-8432-4

Index

O. Gazi, *Information Theory for Electrical Engineers*, Signals and Communication Technology, https://doi.org/10.1007/978-981-10-8432-4

Printed in the United States
By Bookmasters